9/87

D1294544

THE BIRTH OF THE EARTH
A *Wanderlied* Through Space, Time, and the Human Imagination

David E. Fisher
Previous Books

NOVELS
Katie's Terror
Variation On A Theme
The Last Flying Tiger
A Fearful Symmetry
Crisis

NONFICTION (Children and Young Adult)
The Third Experiment
The Ideas Of Einstein
The Creation Of Atoms And Stars
The Creation Of The Universe

THE BIRTH
OF
THE EARTH

A *Wanderlied* Through
Space, Time,
and the Human Imagination

DAVID E. FISHER

New York COLUMBIA UNIVERSITY PRESS *1987*

James Stephens' poem, "What Tomas Said in a Pub," was
published by Macmillan in 1909.

Library of Congress Cataloging-in-Publication Data

Fisher, David E., 1942–
The birth of the earth.

Bibliography: p.
Includes index.
1. Solar system.   2. Cosmology.   I. Title.
QB501.F53   1987   523.2   86-14037
ISBN 0-231-06042-4

Book design by Laiying Chong

Columbia University Press
New York   Guildford, Surrey
Copyright © 1987 Columbia University Press
All rights reserved

Printed in the United States of America

This book is Smyth-sewn.

For Hans Frese
Professor of German,
Trinity College, 1952/3

and

for my parents
Henry R. and Grace S. Fisher

CONTENTS

ACKNOWLEDGMENTS

I am grateful to the following for assistance in various stages of the manuscript, ranging from quick answers to long discussions and from published papers to personal notes and reminiscences: Ray Reynolds, David Black, and Jeffrey Cuzzi of NASA; John Reynolds of Berkeley; Bob Clayton and Lawrence Grossman of Chicago; M. M. Woolfson of York; Ron Fisher of Baylor; and William Fowler of Cal Tech. Without their help there would have been many more errors in the book; it goes without saying that the errors remaining are my responsibility alone. I am also indebted to Kay Hale and Sara Jeffreys of the University of Miami RSMAS Library for continual outsearches.

WHAT TOMAS SAID IN A PUB

I saw God! Do you doubt it?
Do you dare to doubt it?
I saw the Almighty Man! His hand
Was resting on a mountain! And
He looked upon the World, and all about it:
I saw Him plainer than you see me now
—You mustn't doubt it!

He was not satisfied!
His look was all dissatisfied!
His beard swung on a wind, far out of sight,
Behind the world's curve! And there was light
Most fearful from His forehead! And He sighed—
—That star went always wrong, and from the start
I was dissatisfied!—

He lifted up His hand!
I say He heaved a dreadful hand
Over the spinning earth! Then I said,—Stay,
You must not strike it, God! I'm in the way!
And I will never move from where I stand!—
He said,—Dear child, I feared that you were dead.—
. . . And stayed his hand.

—James Stephens (1880–1950).

THE BIRTH OF THE EARTH
A *Wanderlied* Through Space, Time, and
the Human Imagination

INTRODUCTION

Stories told of the history of science all too often portray a smoothly rolling film sequence of discovery, refutation, modification, and final verification. Someone conceives of a scenario to explain a concept such as our existence, and others find faults with it; point by point the arguments are attacked and experimentally or theoretically they are demolished, modified, or established. The scientific process thus described is a lovely fairy tale, but in my own experience I have found it to be usually more random and chaotic, explicable as much in terms of fashion and passion as in those of logic and measurement. Points of argument are often ignored, points of view skirted around; various counterarguments are presented simultaneously so that future impetus is split into skewed directions, nevermore to meet again; theories as well as experimental data are forgotten, resurrected, rediscovered, reforgotten; they are proven wrong one year, accepted the next, discarded without argument the third. And somehow, out of all this chaos, we slowly learn. It is not a calm, reasoned, logical learning experience: but it's not only the best we have, it's the only game in town.

The historian of science as well as of other torturous human paths cannot tell a story without inserting a thread through the holes to hold it together, to make a pleasing or at least an organized quilt out of the various ill-fitting patches. To tell it as it truly occurred would be bewildering, but to tell it as a coherent story is often untrue. So a compromise must be reached, in which we pick and choose among the refuse of history to find a scheme which approximates both the truth and a story with a beginning, a middle, and an end, and finally then we say, "This is the way it happened."

Right, then. This is the way it happened.

CHAPTER ONE

•

IN THE BEGINNING

In the Beginning God created the Heavens and the Earth.
Well, not exactly.

THAT FIRST SENTENCE represents an astonishing leap of human
imagination. It replaced a universe of chaos with one of order. The
universe—and the day-to-day fortunes of humanity—were no longer at
the mercy of the whimsies of innumberable gods known and unknown,
but were created by and subject to the one overriding purpose and stern
discipline of a just (if occasional vengeful) God. The sun did not rise in
the morning, bringing the light and warmth necessary for our survival
because the god Shamash felt like driving his chariot across the sky, but
because the sun was created by the Lord aeons ago precisely to bring us
light by day; it existed for that purpose and would never stop, would never
fail to appear. It was no longer necessary to worry each night whether
Shamash might change his mind the next morning and fail to appear; there
was no need to cut out the gizzards of chickens or the hearts of virgins
and spread them on altars to induce Ushas to bring the dawn or Enki to
send the fish for supper or Immer to water the fields with rain: the universe
had an order, a discipline, a purpose, and so did we. That first sentence
established the foundation of an understandable universe and became the
basis of a system of moral and religious values that has lasted thousands
of years and, though observed more in the breach than in the practice,
still underlies the fabric of our civilization.

The second sentence is even more astonishing. It became possible to formulate it only in the second half of this present century: incredibly, we can now say with the utmost certainty something about the creation of our world. We know quite clearly that the Earth was not created *in the Beginning* at all: many billions of years actually elapsed between the creation of the first heavens and that of the Earth, many stars were born and passed into oblivion before ever the Earth was even a mote of dust in its Creator's eye, to coin a symbolic phrase.

The evidence for this comes from several different lines of scientific enquiry, from nuclear physics and geology as well as astronomy, and all the evidence fits together to form a proof beyond all reasonable doubt. We know, to begin with, that the universe we live in was created in a fantasic explosion we call the Big Bang. We don't know what happened or existed before this event, whether other universes existed in an unending chain beyond the beginning of time and whether the process will continue infinitely far into the future, or whether our present universe is the sum total of existence. Our imaginations boggle and collapse under the weight of such heavy questions. But we do know that our present universe was created in that moment of the Bang: we see the evidence of that event in the nearly homogeneous background radiation that now pervades the universe, the slowly dying relic of that first radiation flash. We see the evidence also in the motions of the galaxies, which are still being blown away from us and from each other with the force of that initial explosion. This latter observation, that of the motion of the galaxies, was the first hint we ever had of the overall structure of the universe. It depends on measurements of the spectra of wave lengths of light emitted by hot gases.

It was discovered early in this century that when a gas is heated to incandescence the light it emits consists of a series of discrete wavelengths which are typical of the type of gas; in fact, the spectrum of wavelenghts provides a spectroscopic fingerprint by which the identity of the gas is revealed. In this way astronomers analyzed the light coming to us from the stars in our galaxy, and found that all the stars were composed overwhelmingly of hydrogen. When they looked at the spectra of light coming from other galaxies, however, it was subtly different: it showed the characteristic relative spectrum of hydrogen, but the absolute values were always shifted to higher wavelenghts. This became explicable in terms of Einstein's General Theory of Relativity and an experimental observation known as the Doppler Shift.

In 1917, even before his theory of general relativity had been proven correct, Einstein attempted to apply it to the entire universe. He managed to find a unique solution to the relativistic equations which specified a homogeneous universe with no motion, space curved and without limit, yet finite, and with time uniform but infinite. He was mildly disturbed when the Dutch mathematician Willem deSitter found another solution in which the universe was empty but had the peculiar quality that if any small amounts of matter were introduced into it they flew apart spontaneously and continued to recede into infinity. Since, however, our universe was demonstrably not empty, it was possible to dismiss the DeSitter solution as irrelevant.

Nothing, however, is irrelevant to mathematicians, and by 1924 the Russian mathematician Alexander Friedmann had discovered a whole spectrum of possible solutions in which matter, which here was as natural a component as in the Einstein universe, spontaneously flew apart as in the deSitter universe. This model was extrapolated backward in time by the Belgian Abbe Georges Lemaitre, to a point of infinite density at zero time known as a "singularity." This word is used to denote a situation that is physically impossible—that is, impossible within our laws of physics. It arises mathematically as a function that is not well-behaved: noncontinuous, with a noncontinuous derivative.* An example might be something like a radar plot of a jet airplane's trail from Miami to New York which instantaneously becomes zero over Richmond and then just as abruptly reverts to its proper value again. This would be impossible as a plot of a real airplane in our real world; if it showed up this way on a radar screen the operator would conclude that the system was malfunctioning. And that is what Einstein and many others thought at first when the Friedmann/ Lemaitre solutions to the relativistic equations showed a singularity at the beginning of time.

*To be more precise, if $f(z) = u(x,y) + iv(x,y)$ and if u and v and their partial derivatives with respect to x and y are continuous and satisfy the Cauchy-Riemann conditions

$$\partial u/\partial x = \partial v/\partial y \text{ and } \partial v/\partial x = -\partial u/\partial y$$

in a given region, then $f(z)$ is said to be *analytic*. A *singularity* is a point at which $f(z)$ is *not* analytic. "Real" functions are analytic everywhere under conditions we consider normal in our universe.

The flinging apart of matter in these solutions was also disturbing; it meant that the universe could not exist as a stable, static system. Rather it had to be continually expanding. (Actually, as I mentioned, the Friedmann solutions are an entire family of possibilities, including the possibility of contraction as well as expansion; but certainly a static, nonmoving, time-independent universe is not one of them.)

These two problems suddenly became the solution known as the Einstein-Friedmann universe when it was realized what they mean: that the universe began as a singularity, in a state which does not correspond to any aspect of physical reality today, a state of infinite compression and density which *in the Beginning* exploded and sent all the matter in the universe spinning outwards. Today that matter, in the form of galaxies, is still spinning out, expanding, receding from itself.

We see this expansion of the universe in the Doppler Effect. The shift in wavelength of the lines of the hydrogen spectrum is due to the motion of the light source—the distant galaxies. When an object emitting light waves is moving toward the observer, the wavelength of the light appears to him to be shortened; when the object is moving away, the wavelength appears lengthened. This effect was first discovered by the Austrian scientist Christian Johann Doppler, who thought that observations of starlight would show random motions of the stars: some moving toward us, some away from us. Within our galaxy, such motions are so small as to be all but indiscernible, but the light from other galaxies all show a shift to longer wavelengths: every galaxy is moving away from us and away from each other. Not only that, but the further ones are receding at faster velocities, proportionally to their distances. This is precisely the effect to be seen as the aftermath of an explosion, and so the observations together with the theory tell us clearly the story of the Big Bang.

They tell us more: they tell us *when* it happened. Simply by taking the measured distances of the various galaxies together with their measured velocities, we can tell how long it took them to get where they are; the calculation is simply the inverse of determining how far an airplane has traveled, from a knowledge of its take-off time and its speed. Unfortunately, it's not quite so simple: there are large errors in our measurement of the distance of the galaxies. But taking these into account, we can place the time of the Big Bang at certainly within 12 to 40 billion years ago, and probably within 15 to 20 billion. In other words, the universe began at least 12 billion years ago.

That's the evidence from astronomy. Nuclear physics tells us much the same thing. We know today that of all the different elements present on earth only hydrogen and helium were created during the Big Bang. The other elements—including oxygen, silicon, iron, uranium, and all the various elements which form our planet and our bodies—were created by nuclear reactions occurring in the interiors of stars scattered through the galaxy during the billions of years before the sun and earth were formed.

The realization of these processes began in the war year 1944 when Fred Hoyle (then a young scientist working on radar for the British Royal Navy, now Sir Fred Hoyle of Cambridge and California and Wales, arguably the greatest astrophysicist of his generation), took time out from a visit to the U.S. Naval radar installations at San Diego to visit nearby CalTech and talk about nuclear reactions. The source of energy in stars like the sun had only recently been understood: in 1929 the Danzigian nuclear physicist Fritz Houtermans and the British astronomer Robert Atkinson applied the Russian George Gamow's theory of artificial radioactivity to thermonuclear reactions and concluded that such reactions could produce the solar energy, and one decade later the then German, now American physicist Hans Bethe worked out the details of the particular reactions taking place at the extremely high temperatures in stellar cores. The basis for this is the conversion of four hydrogen nuclei into one helium. The mass of the helium is slightly less than that of the four hydrogens, and the mass difference is converted into energy according to $E = mc^2$.

Hoyle was intrigued not only with the energy production, but with the fact that one chemical element was converted into another: alchemy had been removed from the realm of medieval fantasy and had been transplanted whole and hearty into the center of stars. He began to think about what further nuclear reactions might take place later in the star's lifetime, converting helium successively into heavier elements. But he got sidetracked on other problems, and it was Ed Salpeter of Cornell who discovered that helium could be thermonuclearly fused into carbon by the nearly simultaneously coming together of three helium nuclei. Hoyle immediately jumped back into the problem, calculating further that carbon would fuse with another helium nucleus to produce oxygen. In fact, he found that this process would be even too effective, removing all the carbon as soon as it was formed. The result would be a universe without carbon, and that doesn't happen to be the universe we live in.

The only way to save the situation and preserve a universe built by stellar nucleosynthesis was if the production of carbon was more effective than Salpeter had originally calculated. After fiddling with the mathematics a bit, Hoyle found that the process would be possible if the carbon nucleus had a metastable state at a particular energy level, 7.65 Mev, which would have the effect of drawing its formation reaction forward.

The consensus during those years was that although nuclear reactions in stars were the source of stellar energies, the universe had been created with all the variety of elements already present, or at least they had been created in the first few moments of the universe; the conversion of hydrogen into helium was thought to be the only significant alchemical reaction occurring today. Then in 1953, on another visit to CalTech, Hoyle discussed the problem of the creation of carbon with Willy Fowler, a nuclear physicist there. Hoyle insisted that the 7.65 Mev state of carbon-12 must exist. Previous experimental work had in fact found some doubtful evidence for such a state, but later and more careful research had not seen it at all. When Hoyle argued that he was convinced it was there and that the experiments should be redone even more carefully, Fowler's initial reaction was, he later recalled, "Go away, Hoyle, don't bother me." But Hoyle was a persuasive speaker and there were intelligent listeners at CalTech. A research group carried out the definitive experiment, in which the predicted state of carbon-12 was found precisely where Hoyle had predicted it must be, and Fowler became a believer.

A couple of years later he spent his sabbatical year at Cambridge, where he was waylaid by the husband and wife astronomy team of Geoffrey and Margaret Burbidge, who wanted the help of a nuclear physicist. Inspired by Fred Hoyle's work, they were investigating the possibility of element production in stars via neutron-induced reactions. Hoyle was also in residence in Cambridge, and the four of them worked together that year, often spending evenings over their calculations in the Burbidge's tiny flat on Botolph Lane, "a flat with a roof that leaked in the nearly incessant winter rain, and to which the only access was an unlit passage past innumerable outhouses and up a very narrow, winding staircase." When Fowler's sabbatical was done he arranged for the Burbidges to follow him back to California; in 1956 Fred Hoyle also accepted a professorship there, and the four of them produced the bible of nuclear astrophysics, "Synthesis of the Elements in Stars," known to succeeding generations of scientists colloquially as B²FH, after the authors. In this paper they presented in whole cloth an understandable and workable system of a variety of nuclear reactions in different stellar conditions that would in toto produce the elements seen in the universe in their proper proportions. In 1983 Fowler was awarded a Nobel Prize for this work, invoking surprise in several quarters that the others were not similarly honored.

The models of stellar nucleosynthesis which began with the Fowler/Hoyle work tell us how the different isotopes of the various elements are created. One such pair consists of uranium-235 and uranium-238. We know that they must have been created in roughly equal abundances, yet the ratio of U-235/U-238 today on earth is only 0.0073. What happened to all the original U-235? Well, the answer is clear: U-235 is radioactive, with a half-life of 713 million years; every 713 million years, half the U-235 is transformed into other elements. U-238 is also radioactive but its half-life is much longer so that every 713 million years only 10 percent of it changes to other elements. Therefore the ratio U-235/U-238 is always decreasing at an easily measurable rate, and it is a simple matter to reverse the calculation and determine how long ago the ratio was unity; the answer is about 6 billion years ago. This tells us that the isotopes of uranium were created at *least* as long ago as that: the true age is actually longer since they (and all the elements) were created over a period of aeons of time in a succession of stars. The calculations become more complex and model-dependent, but using several pairs of radioactive isotopes of different elements the answer always comes out to be the same, giving a rough date of 12 billion years and a firm lower limit of 9 billion years. Since these elements were created in various stars within our galaxy, the result is an average age for the galaxy. This fits perfectly with our Big Bang model in which the universe itself is something more than 12 billion years old, most probably 15 to 20.

Now what about the age of the earth? The answer comes again from nuclear physics, this time combined with geology. Radioactive elements behave chemically just like normal stable elements. In particular, they combine naturally to become part of the minerals and rocks on earth. The difference is that, whether combined in a rock structure or not, wherever they are and in whatever form, they continually decay and change into other elements at an unalterable, predetermined rate. Obviously the ratio of a parent radioactive element to its daughter is a continually changing function of time. Simply by measuring this ratio one can determine the age of the rock, and the age of the earth is to a first approximation equal to that of the oldest rock measured. Better approximations can be obtained by particular methods on particular geological samples, as will be discussed in chapter 7. Many such measurements in the past fifty years have shown that rocks on earth commonly reach ages of millions, hundreds of millions, even billions of years, providing the first conclusive evidence against the seventeenth-century biblical-based estimates of some literal-minded Christian theologians, which gave results of a few thousand years—but the *oldest* age of any rock on earth is only a few billion years, less than half the 9 to 12 billion-year age of the galaxy.

By measuring the isotopic composition of radiogenic and primordial lead averaged for the whole earth, it is possible to show that the earth is just about 4.5 billion years old. Incredibly enough, similar experiments on meteorites show the same result, which means that they and the earth were created together, within a possible error of only 100 million years.

Let's stretch the error estimates as much as we possibly can; still the earth is unquestionably less than 5 billion years old. And the galaxy is more than 9 billion years old, and the universe itself is more than 12 billion years old.

And so in the Beginning the heavens were created, but the earth came a long time later, many billions of years later. Many billions of stars lived and died, creating the elements that now form the earth and spitting them out into the vast interstellar spaces, before finally the clouds of dust and gas began to fall together to form our sun, our solar system, our earth, and ourselves.

The story of how that all happened is an incredible one, but just as incredible is the story of how we dug out the secrets of our own creation, piece by piece, and put them together. That's what this book is about.

SOURCES: Burbidge and Burbidge 1982; Hoffman 1972; Hoyle 1982; Lovell 1981

CHAPTER TWO

•

THE SPINNING EARTH

THE CONCEPT OF One God was the first major step man took in bringing order to the universe, in understanding that the world we live in—bewildering as it seems to be—is in fact governed by strict laws. The realization that such laws do exist and are discoverable, knowable, and independent of the moods of any Creator did not come for more than another two thousand years, not till the middle of the seventeenth century when Kepler propounded his laws of planetary motion and Newton explained them with his laws of gravity and inertia.

Today we take the laws of physics for granted. They have become part of our everyday lives and we neither wonder at them nor doubt them, nor do we realize that they are statements of ignorance as well as triumphs of the human intellect.

Perhaps the most basic of these, for example, is the law of conservation of mass-energy, which states that mass-energy can be neither created nor destroyed. Ignoring relativistic considerations which merge the previously separate categories of mass and energy in a manner now commonly accepted though difficult to understand, we speak of the conservation of mass or of energy. The statements are simple and intuitively obvious: matter does not disappear before our eyes nor does it come into being. Work must be expended in order to generate energy: there is no free lunch.

But why? Where is it written that this must be so? Why could not the universe exist with atoms spontaneously appearing and disappearing, and with energy freely available? In 1948, for example, the Steady State theory of the universe was proposed, incorporating the idea that atoms were continually created out of nothing. Nearly three decades later the theory was proved wrong, but it was new experimental data rather than the concept of atomic creations that did her in. In fact if the law of mass-energy conservation is truly a universal law for all time, then where did all the mass and energy in the universe come from?

We don't know the answers to these questions. Or rather, by the very definition of the meaning of a scientific law and the meaning of the word *why* the questions become internally contradictory, mutually exclusive. To ask why something is true is to ask that an explanation be derived from a simpler, more basic principle: Why is 2 + 2 equal to 4? Because 1 + 1 is equal to 2, and (1 + 1) + (1 + 1) equals 4. But what we *define* as scientific laws are ultimate and basic principles: if we could derive them, then the principles from which they were derived would replace them as our laws.

For example, what makes the earth go round?

When man first realized that the sun, moon, and stars rise and set every day and night in the same way, he asked why? And he answered with a set of mythological stories. Today we accept without question the dramatically unobvious notion that the rising and setting of the heavenly bodies is actually a manifestation of the fact that the earth is spinning on its axis, rendering the ancient explanations irrelevant and unnecessary. But we may now quite reasonably ask instead, why should the earth spin? The biblical interpretation that it does so in order that the sun might bring us alternating day and night is no longer satisfying, since that answer rests on the assumption that the universe was created for our convenience; this notion was acceptable when we thought of the universe in an Aristotelian sense, with the earth at the center of it all, but it became obviously ludicrous when we realized the immensity of the real universe and the insignificant location in it of the solar system we inhabit, and indeed of our own location in this one particular solar system. We are a natural part of the universe, but certainly not its center nor its *raison d'etre*.

We might improve on the biblical interpretation by realizing from observation that every other planet we see is also spinning; it would then be a reasonable step to conclude that it is a natural law of the universe that all planets spin. The earth is a planet, *ergo sum:* the question is answered.

This proposed law—that all planets spin—would reveal both knowl-

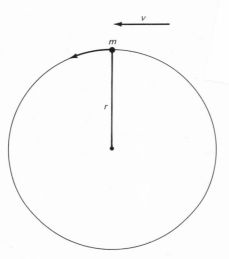

Figure 2.1.

edge and ignorance of the most profound sort. It would incorporate the empirical knowledge that all planets are spinning, and to have learned this—to have learned something of such basic significance about objects hundreds of millions of miles away, infinitely far away to us creatures locked on our own planet—is a magnificent accomplishment. And yet this law would also be a confession of ignorance, for it would reveal that we have no idea as to *why* they are spinning. We would be saying that we know they spin and that is the extent of our knowledge.

But in fact we know a bit more than this. We can indeed derive this "law of spinning planets" from a more basic law: the law of conservation of angular momentum, illustrated in figure 2.1.

Figure 2.1 shows a body of mass m moving around a center O in a circle of radius r with a velocity v. It represents any number of situations: if the line r is a string and the mass m is a stone and the center O is a person, it represents someone swinging a stone around his head on a string. If O is the sun and m is the earth, and the force holding m to O is gravity rather than a string, it represents the earth revolving around the sun. If O is the center of the earth and m any point on the surface of the earth, it represents the earth spinning on its axis.

All these situations have something in common: if no energy is added or subtracted from the system, the product of m x r x v will remain constant. The product of these three numbers is named the *angular momentum,* and the previous sentence comprises the Law of Conservation of Angular Momentum. First we'll discuss it, then try to explain it.

If the string holding the stone is pulled in, the distance r gets smaller; then in order for the angular momentum to remain unchanged, either m or v must increase. Since mass conservation prevents the stone from getting bigger, the velocity must: the stone will whirl faster. Conversely, if some string is let out so that r increases, the stone will whirl more slowly. All this is without the person whirling the stone making any effort to spin it faster or slower.

The effect is seen commonly on television whenever figure skating on ice is shown. A skater goes into a whirl with arms outstretched, then pulls her arms in across her chest *and she spins faster.* This is because the radius around her spin axis, from the center of her body to her outstretched fingertips, decreases when her arms come in close; since her mass can't increase, her spin velocity must. To stop, she'll throw her arms wide again: the resulting increase in r makes her v decrease drastically and she can easily stop without falling over.

During the formation of the solar system, the earth as it formed acquired a certain amount of spin angular momentum both from the tumbling motion of its component parts and from late collisions with asteroidal-type objects, all of which will be discussed in detail later in this book. The point now is that once the earth acquires this angular momentum, it can't lose it—and so it keeps on spinning, aeon after aeon, until the last syllable of recorded time.*

So why does the earth spin? Why do all planets spin? Because during their formation they acquire some angular momentum, and because of the law of conservation of angular momentum they can't lose it, and so they spin.

But why must angular momentum be conserved? No one knows. The law is, at the present time, the ultimate expression of our most basic knowledge about rotating objects; we can neither derive it nor explain it

---

*In this discussion I have treated each planet as if it were isolated in space, ignoring higher-order coupling effects which would only cloud the basic concept of angular momentum.

by more basic principles. We have to be satisfied if we can explain the observed universe with a set of such basic principles. This is at the same time our glory and our limitation; we have discovered that the universe is an orderly one governed by law, but the ultimate reasons for these laws lie in that undiscovered country whose bourne none of us travelers have yet reached.

SOURCE: Morrison 1971

CHAPTER THREE

•

THE EARTH STANDS ALONE

THE INEXORABLE PROGRESS of human comprehension of the universe is marked by occasional, sudden leaps in understanding—which sometimes turn out to be *mis*understandings, and yet which carry us forward toward the truth. In general philosophic terms the great advancements in our *Weltbild* were, first, the prehistoric concept that the world was not totally irrational but was ruled by a host of gods who could be pacified and thus, in a sense, the world's terrors could be palliated; second, the Jewish concept that there was just the one god whose scheme rested on moral order; and third, the concept of a universe ruled by nonanthropomorphic laws. The first two of these great advancements turned out to be simply wrong—at least there are few people today who would argue for the first one, and I suppose we'd better leave it at that—but they took us forward on the road to whatever ultimate truth may exist. They established, necessarily piecemeal, the concept that there is an order in the universe; a concept much more difficult to assimilate than that of general relativity, a concept that took us thousands of years to appreciate.

Today our advances in science are held more closely to the path of truth by our insistence that they be tested observationally and modified or discarded when necessary. It must have been obvious to our remote ancestors that pouring chicken blood on the ground in arcane symbolic designs did not really improve the next year's harvest, and yet they clung to their beliefs generation after generation: because the only alternative was an irrational, whimsical world over which they would have had no control at all. It's more comforting to hold the reins in your hands even though they dangle loose and unconnected at the other end than to throw them away altogether and ride screaming and helpless into that good night.

The first beginnings of a scientific description of the universe, based on observations and tested and modified by them, were also totally wrong. They were not only wrong, they led to a dead end; and that is the worst scientific sin of all. If you take a wrong turn when you're driving toward a destination, you'll generally come out *somewhere,* and if you have any sense of direction you will have made at least a little progress toward your destination—you hardly ever get yourself so turned around that you head backward. But if the turn leads to a dead end, you just have to turn around and go back and start over.

The Greek concept of a geocentric universe with the heavenly bodies revolving around it in perfect spheres was just such a dead end. The Greeks, in fact, never came up with any sort of theory about the origin of the universe: their concept was of such a stable, static, perfect universe that it never occurred to them to think about a beginning or an end. Their idea was not only wrong, it was sterile.

But it was intelligent and reasonable, and it was based on careful observations and modified accordingly as the observations became more accurate, and it lasted almost without argument for more than two thousand years; for all of these reasons it is worth talking about.

The basic observation was that the sun, moon, and stars all rose every day (or night) in the east and sunk in the west. In the sixth century B.C. the Pythagoreans popularized the idea that while everything here on earth changes and decays, the stars are unchanging through time immemorial, and therefore perfect. They concluded that their motion must reflect this perfection, and must therefore be circular: the circle, without beginning or end or angles or capricious change of direction, is obviously the essence of perfection. And this concept was borne out by observation: if you look at the sun and the stars you can see them moving around the earth in perfect circles—at least this made more sense than thinking of the sun as being created anew every morning and crashing to destruction every night. It had the further advantage of removing the fear that one morning a capricious Helios might decide to change his ways and doom all of us to everlasting darkness.

Over the course of the next three centuries a rational cosmology was

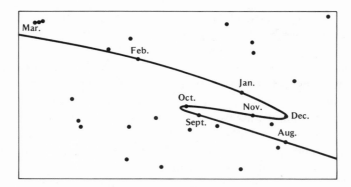

Figure 3.1.

built on the observations of the heavenly bodies, and on the theoretical basis of the perfection of the circle. It was a grand theory, simple and beautiful. Unfortunately, people then discovered the curious motion of the planets. Unlike the fixed stars which rose every night in exactly the same positions relative to each other and so could easily be visualized as fixed in an eternal sphere, these wandering stars moved apparently at random throughout the night skies. Centuries of careful observations by Greeks and Babylonians unknown slowly gave a pattern to these motions, but a pattern difficult to fit within the concept of spherical revolutions around the earth. It was evidently Plato in the fourth century B.C. who placed importance on the solution to this problem, throwing it out to his students as a university professor today might suggest a Ph.D. thesis topic. It was not an easy problem. There were seven recognized "planets": today we call these bodies the sun, moon, and the five planets Mercury, Venus, Mars, Jupiter, and Saturn. Together they all move westward every night (or day), but gradually—night by night—they slip eastward against the background of the fixed stars, and at different rates so that, for example, Mercury and Venus complete one revolution against the stars in one year while it takes Saturn nearly thirty years. In addition, the five latter planets exhibit "retrograde motion": as they come moving night by night across the sphere of the fixed stars, they suddenly turn around and move *backward* for varying periods; then once again they take up their "normal" direct motion, as shown in figure 3.1. Mercury begins to move backward every 116 nights while Mars stays on its normal course for 780 nights before turning around.

Mercury and Venus, in addition, always stay close to the sun so that they are seen only just before dawn or just after dusk. In fact, it took a long time before observers realized that the dawn planets were the same objects as the dusk ones.

Plato's suggested "Ph.D. thesis" was to account for these planetary motions, which were thus both erratic in regard to the celestial stars and yet secularly regular, within a scheme of circular motion around the earth as center. His pupil Eudoxus of Cnidus (408–355 B.C.) was the first to succeed, providing a complex mathematical model in which the seven planets moved according to the motions of twenty-seven spheres which revolved independently, their collective motions giving the final observed planetary orbits. During the next hundred years further observations showed discrepancies between the model and the actual motions, despite revisions by Aristotle, until Appolonius of Perga (250–220 B.C.) introduced the concept of "epicycles." This was the idea that each planet actually moved in a circle whose center itself moved in a circle around the earth. Hipparchus (c.130 B.C.) moved the earth from the center of the circles described by the sun and moon, in order to account for their apparently changing velocities as observed from earth; this left the sun and moon revolving around an empty point, but that didn't bother anyone since Appolonius' epicycles had already established the precedent. Finally, by about A.D. 90, Ptolemy of Alexandria encoded all these ideas in his *Almagest,* a massive book which described the motions of all the bodies in the universe. A simplified diagram is shown in figure 3.2.

By imposing epicycles upon epicycles, it was possible to fit the known motions of all the planets exactly. The theory therefore accounted satisfactorily for all the facts within a scheme of "simple" circular motion; simplicity, after all, is in the eye of the beholder, and satisfaction is to be found in his heart.

Everyone was satisfied. Nearly everyone, that is. Aristarchus of Samos (c.270 B.C.) and Seleucus of Babylon in the following century argued for a universe with the sun at the center and earth revolving around it, but nobody took them seriously. There was no reason to. The heliocentric universe explained nothing the geocentric universe did not. It had no great advantage to balance against its overwhelming disadvantage: it postulated an earth that moved, whereas clearly anyone sitting on this earth could feel that it did not.

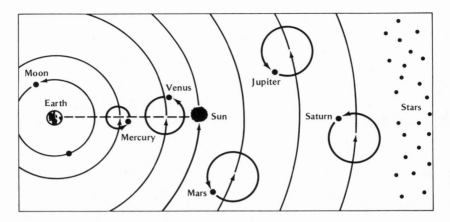

Figure 3.2.

The concept of a spherical earth, on the other hand, was firmly established in the Ptolemaic system and didn't bother anyone. When it had first been proposed, centuries before, it had been a particular sore point: if the earth were spherical, people on the bottom would fall off. In fact, those on the top would tend to slip off, wouldn't they? The more theoretically minded argued that, willy-nilly, the earth *must* be spherical because of the perfection of the sphere and a world not perfect would hardly be worth living in.

Such an argument has a certain attraction but the real breakthrough came with the concept of natural motions, advanced by person or persons unknown sometime previous to the third century B.C. This idea pointed out that it is only an illusion that things appear to fall down: we take for granted that *down* is a preferred direction, but we are mistaken. Instead the preferred universal vector is radially inward toward the center of the earth, and the natural motion of all things on earth must therefore be toward this center. Throw a stone or a person off a cliff, freeing it from all earthly restraint, and its natural motion will take it toward the earth's center. Anyone watching will think the stone/person is falling *down,* but that's only because we don't see the large picture: in our limited view, the radial vector looks like down, but it will look the same to people or stones anywhere on this spherical earth.

This idea did not lead to the concept of gravity. There was no force

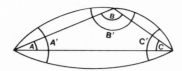

Figure 3.3.

implied; indeed the concept was that any object freed of all forces would take up this natural motion toward the center. It found a corollary in the natural motion of the heavens, and answered a perplexing question: why did the stars and planets move in their circular paths? What pushed them along? The answer was that no push was necessary since the natural motion of the celestial spheres was circular, just as the natural motion on earth was radially inward.

By the time of Ptolemy there was in fact experimental evidence that this view was correct. The Egyptians had long been expert in surveying skills, necessary because of the great size of the estates necessary to bring in an economically sufficient harvest in that region. While laying out triangular areas which extended many miles they had come upon a curious fact: the sum of the angles enclosed by these triangles always exceeded 180 degrees. But Euclid (c.300 B.C.) had convinced everyone that the sum of the angles of a triangle must always be exactly 180 degrees.

The obvious explanation at first was that the land measurements were in error, but careful repetition over the years always gave the same result. The answer came by considering what happens if the triangle is laid out not on a flat surface, but on the surface of a sphere: it is distorted, swollen outward (figure 3.3).

The spherical angle A' is greater than the flat angle A, and the same is true of the other two angles, with the necessary result that the sum of the angles A'B'C' must be larger than the sum of ABC, and so greater than 180°. The measurements are therefore explained if and only if the apparently flat land of Egypt is actually curved like the surface of a sphere.

The first actual measurement of the earth's curvature was made by an Egyptian Greek named Eratosthenes (276–195 B.C.), a native of the southern Egyptian city of Syene who became royal librarian at the Grand Library of Alexandria under Ptolemy Euergetes. He had noticed without paying any particular attention while growing up in Syene that on every June 21

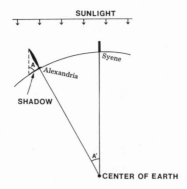

Figure 3.4.

at noon the sun was directly overhead: it cast no shadow from a vertical tree. But now at Alexandria he realized that this was no longer true, and in a flash of genius saw the reason why—the curvature of the earth. Alexandria lay 480 miles due north of Syene, and so when the sun is directly over Syene it must lie at an angle over Alexandria. Not only that, but the angle could be measured by the shadow cast by a vertical tree or pillar in Alexandria at exactly that time, as shown in figure 3.4.

The angle A is determined from the height of the tree or pillar and the measured length of the shadow at Alexandria, at a time when there is no shadow cast at Syene (noon on June 21). Alternate interior angles of parallel lines are equal, and so A' is equal to A. The measured value was 7°, and so an angle at the earth's center of 7° defines a surface chord of 480 miles. Since there are a total of 360° in any circle, the circumference of the earth had to be 24,000 miles and its radius $24000/2\pi$, or just under 4,000 miles. The correct values are 24,989 miles for the circumference and 3,950 miles for the radius. (There is a bit of controversy over the accuracy of his measurement, since he used the Greek unit of *stadia* instead of the modern mile, and there is some uncertainty as to the exact length of the stadium, but this is our best estimate. Without nit-picking, his method was ingenious and his data wonderfully accurate.)

It's interesting that this early Greek knowledge of the earth was not entirely but to a large extent lost in the succeeding centuries of European barbarism. When Columbus embarked in 1492 he knew that the earth was spherical, but was seriously wrong in estimating its size. Science had been replaced largely by religion and Greek by Latin, so the size of the earth was newly redetermined by such people as Cardinal Pierre d'Ailly, Chancellor of the University of Paris at the turn of the fourteenth century, who instead of measuring shadows on the earth's surface went back to the original source of all knowledge: in the biblical Apocrypha he found the definite statement that one seventh of the earth's surface is covered with water. Knowing (wrongly) the area covered by land, he calculated that the (one) ocean separating Europe from China covered only about four thousand miles.

Such information was the basis for Columbus' voyage in search of a westward route to India. Had he known the true distance he would have known that his ships and supplies were incapable of making it, and he never would have tried. On such grand bases is history made.

SOURCE: Morrison 1971

CHAPTER FOUR

•

NEVERTHELESS, IT MOVES

W HEN THE ROMAN EMPIRE stretched its slow thighs and
slouched toward the Mediterranean to be born, it marked the end
of civilization in a very real sense. For nearly two thousand years afterward
every form of intellectual endeavor plummeted as the Romans overwhelmed the Greeks and then the barbarians overwhelmed the Romans.
Today we still study with awe the Greek accomplishments in astronomy
and mathematics. We still today sit entranced and emotionally rocked at
the power and beauty of Greek drama such as *Oedipus Rex* and *Elektra*.
But when was the last time you saw a Roman play? The Pax Romana stifled
as well as pacified, and civilization blinked sleepily and very nearly went
out; certainly it went into a very long sleep.

The tenders of that sleep were the Arab world, which preserved the
forgotten Greek books through the dark anarchy that followed the Roman
Empire. In the twelfth and thirteenth centuries a slowly emerging European
community of scholars, almost all nurtured by the Holy Church, began to
rediscover these works. They could not help being awed by the intellectual
power of Greek science, much as savages in succeeding centuries were
awed by the physical power of the European muskets when those strange
tall-masted ships sailed up out of the mists to their shores. In this case,
however, the savages were made happy by the encounter: the Greek
science jibed satisfactorily with their own religious truths.

But not without a little controversy here and there, a little pushing, a
little fudging. In the twelfth century the influential theological *Sentences*
of Peter Lombard of Paris stated that "just as man is made for God, in order

to serve Him, so the universe is made for man, to serve him. For this reason man was placed in the center of the universe, in order to serve and be served"—a statement which fit perfectly the cosmolology of Plato, Appolonius, Aristarchus, Aristotle, and Ptolemy, and which was now becoming known simply as Aristotelian. But in the following years the Church reacted strongly against the ancient knowledge, rejecting the argument that any knowledge at all was to be found outside the word of God; to seek truth elsewhere was blasphemy, error, and even a sort of intellectual uselessness. For the material structure of the universe was not a fit subject for the minds of serious men: by its intrinsic nature it was necessarily fleeting, unimportant, temporal and temporary, soon to be replaced by the eternal universe of God's heaven on earth which would be brought by the Second Coming. The truly spiritual man would do better to concentrate his intellectual energies on considering the terror and beauty of God's word and on preparing himself for His Messengers. And so the Provincial Council in Paris early in the thirteenth century prohibited the teaching of Greek physics, and shortly after that the Fourth Lateran Council prohibited the Aristotelian cosmology. *Finis!*

The reconciliation of Church and science was due largely to Thomas Aquinas, who lived from 1225 till 1274 and was canonized in 1323. In his *Summa con Gentiles* he argued persuasively for two separate roads to the one great truth. *Reason* is the logical working out of data obtained by the senses, he wrote, while *faith* is the study and acceptance of revelation and authority. He saw no contradiction in these two paths, since all truth is dictated by God who also provided our senses; thus both paths must lead to a knowledge of the universe, i.e., God. Where they appear to contradict each other, faith must be followed, both because of its nature and its source; reason can only *appear* to contradict faith because of the imperfection of our senses and of our reasoning capabilities.

The convoluted reasoning processes by which he managed to reconcile Aristotelian physics and cosmology with the Christian faith are incredible to behold: basically he taught that neither Aristotle nor the Bible necessarily meant what they actually said, but were to be interpreted in marvels of Talmudic twistings and turnings that finally produced, like a magician's pigeon from his wildly flowing colored scarves, a reconciled truth. His arguments were so convincing to the minds of the Middle Ages that the intellectual/ecclesiastical structure he and his disciples erected was elevated to the status of dogma and thus became a caricature of itself: his message that reason and intellect were a valid path to the truth was not itself to be questioned—and without questions there can be no reason, no intellect.

His accomplishment was tremendous, if we ignore the internal contradictions in his logic: he reconciled the laboratory and the Church, the model of the universe predicated on carefully reasoned (if flawed) explanations of careful observations with the model predicated on the untarnishable word of God Himself. But, as did the Roman Empire, his accomplishment overreached itself and ended by propelling mankind on a giant step—backward. He succeeded so well in bringing Aristotle into the Church that the Greek model of the universe, as understood by the Middle Ages, became itself the word of God; to criticize it, to question it, to suggest that it was in essence simply *wrong*, became blasphemy, damned and damnable blasphemy.

In the sixteenth century reason and logic finally began to filter back into the European world, only to encounter this blank wall of absolute conviction. We generally associate this resurrection with the name of Copernicus, but in this we make the same error that the early Europeans made in associating all Greek science with Aristotle, for the workings of a heliocentric universe had been percolating through the medieval world for some time when Copernicus made his appearance on the scene. Despite the political advantages of a Church-science union, error cannot exist forever. The early Greek model with the sun, moon, and stars revolving around a stationary central earth was clear and simple; it *could* have been right. The discovery of the planets wandering through the sphere of the heavens was a damnable complication, but the solution of Appolonius, as modified by Hipparchus, was clever and reasonable; it too *could* have been right. But as more careful observations came to be made throughout the years, this simple scheme of epicycles was shown to not quite fit the motions of the planets; the only way to modify it and still keep the necessary perfection of the circle was to add more epicycles so that the planets moved in epicycles centered on epicycles centered on still more epicycles. Whenever a discrepancy was noted between theory and observation, another epicycle or two was thrown into the model.

Such ad hoc restructuring of a theory is neither unknown today nor necessarily wrong, but the continuing nature of the medieval corrections must give one pause. William of Occam, the "Singular and Invincible Doctor," in 1342 enunciated his famous Razor: "Entities should not be multiplied unnecessarily." The path to truth is most easily followed by keeping it straight; torturous reasoning may be at times necessary and even correct, but always it should arouse suspicion. Simpler solutions should ever be sought, simply because it is all too tempting to cling to a false hypothesis in the face of new and conflicting evidence merely by continuing to modify the hypothesis. Sometimes this can and should be done, but when the modified hypothesis again proves wrong and needs modification, and again and again—it quickly becomes time to cry halt! Better to start over again and seek a simpler solution.

So put the sun at the center. This was a brilliant idea, first publicly proposed by Aristarchus of Samos in about 250 B.C. It was brilliant because it realized that the erratic motions of the planets could be explained as not real but merely apparent if they were being observed from an earth that was itself in motion. It was a brilliant idea but it failed in ancient Greece due to the weight of reasoned scientific notions.

First, it was a common observation that on earth things fall down. Why? With the earth at the center of the universe, this is no problem: the natural tendency of all material objects is to move toward the center. (And then why don't the stars fall down? Ah well, you see, they are quite different, aren't they: aetherial objects they are, whose natural tendency is to move not toward the center of the universe but rather in perfect circles around celestial spheres. This all seemed quite a reasonable explanation.)

But if the earth is in fact not at the center, how could the tendency of earthly things to drop to its surface be explained? Should not earthly objects tend instead to move in circles like the supposed earth? And besides, it was a well-known observation that one could *feel* motion; in particular centrifugal motion, although not yet named, was well known—should not we all fly off the surface of a rapidly revolving earth?

Finally, and most important of all to the scientific mind, Aristarchus' model was shown to conflict with experimental evidence. Around 130 B.C. Hipparchus invented trigonometry, and set out to test the Aristarchian model by experimental observations. If Aristarchus were right, he argued, we would necessarily see the stars in different directions as we moved around the sun in the course of a year. Observations made on a particular star are diagrammed in figure 4.1. In figure 4.1(a) we look at this star from

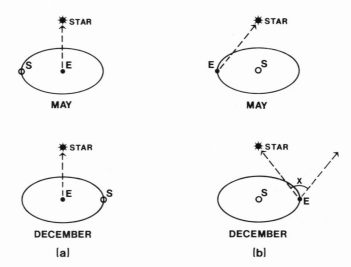

Figure 4.1.

a geocentric earth in May and again in December: although the position of the sun has changed, the direction of the star has not. In figure 4.1(b) we make the same observations from a moving earth: clearly the position of the star as seen from earth has changed, defining an angle X which in fact is a trigonometric measure of the star's distance. Hipparchus made the measurements, and found no observable change in the star's direction throughout the year, corresponding to case (a) and disproving case (b). Therefore the earth does not move around the sun.

The argument was based on brilliant reasoning and careful observations, and it was absolutely convincing for more than 1,500 years. It was wrong, of course, for reasons which we can easily appreciate were beyond the grasp of the ancient world. The angle X is a steadily decreasing function of distance: at very great distances it becomes immeasureably small, i.e., at such distances the observed star will not appear to change direction. With the naked eye the stars do not appear to be significantly further away than the planets, and it was quite reasonable to assume without even thinking about it that the distances to the heavenly spheres are not vast beyond the powers of human imagination. But in fact the angle X is known today to be only about one half of one arc second for even the closest

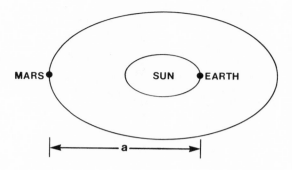

Figure 4.2.

stars, while the accuracy of Aristarchus' observations limited him to the measurement of angles greater than at least 100 arc seconds: he had no chance of seeing the stars change direction through such small angles as they actually subtend.

In fact, no one had the faintest notion of the vast distances to the stars until a Swiss astronomer in the seventeenth century made one of the cleverest calculations in the history of science. It was so clever and remains so unknown that I can't resist making an excursion at this point to describe it. His name was Jean-Phillipe Loys de Cheseaux, and he based his argument on two points: the light we see from an object (a star as well as a lamp) diminishes with increasing distance from us, in fact with the square of the distance; and Mars in conjunction (when furthest away from the Earth) appears about as bright as a typical star. He knew by then that the Earth and planets revolved around the sun, and so he could draw figure 4.2.

The distance from Mars to Earth at its furthest point he calls a. The amount of solar light Mars reflects at that distance is determined by the size of Mars (its area $= \pi b^2$, where b is its radius) and by the sun-Mars distance, and is equal to

$$\text{Mars Light} = (\pi b^2)(L/4\pi B^2),$$

where L is the luminosity of the sun, and B is the distance from Mars to the sun. The light we actually see reflected from Mars is less than this, diminished by our own distance from Mars (a), so the light we see is

Mars Light Seen $= (\pi b^2)(L/4\pi B^2)(1/4\pi a^2)$.

A typical star is at an unknown distance D, and is observed to be just about as bright as Mars. Since all light is diminished by distance by the same amount, the square of the distance, and assuming that all such stars are intinsically about as bright as the sun (fainter only because they are far away from us) Cheseaux said that such starlight must satisfy the equation

$$\frac{L}{4\pi D^2} = \frac{\pi b^2 \times L}{4\pi B^2 \times 4\pi a^2}.$$

And now we see that the unknown luminosity L cancels out of both sides of the equation and we end up with

$$D = 2B \times a/b.$$

Knowing the distances between Earth, Mars, and the sun, and the radius of Mars, Cheseaux calculated that such a typical star is a few light years away from us—something over 20,000,000,000,000 miles away! These were distances that Hipparchus couldn't conceive of, and so he interpreted his measurements to say that the Earth doesn't move, and for 1,500 years no one argued with him.

Finally, slowly, the few people that thought about such things in the medieval world—physicians, priests, astrologers—began to wonder if the universe wasn't unnecessarily complex. King Alfonso X of Castile in the 1200s remarked that if the Lord God had seen fit to consult him before creating the universe, he undoubtedly could have suggested to Him a simpler and more reasonable structure. When even kings begin to have ideas, surely the time is ripe for change!

But the mills of the gods, grinding inexorably, sometimes do grind exceedingly slow. It was a good three hundred years after Alfonso before the simpler solution he had been sure he could find was actually proposed. It began unnoticed by persons today unknown; it was discussed casually, not taken seriously; slowly the discussions began to heat up, simmering and then finally bubbling and percolating and boiling over and through the intellectual community of the sixteenth century.

Nicholas Copernicus was a Polish artist, mathematician, priest, and physician. In the course of his varied studies he traveled throughout Europe, and in Italy picked up the bug of intellectual ferment that questioned everything, in particular the structure of the universe. The basic concept— that the motions of the planets are only apparent because of the motion of the Earth—is so simple that its attraction was overwhelming; he returned to Poland convinced that it must be so. He worked out a mathematically detailed analysis of the idea when he was thirty-six years old, but it wasn't published until he lay on his deathbed thirty-five years later: he was afraid of being ridiculed for proposing such a *different* idea.

Although the mapping out of the heliocentric concept took a good deal of mathematical skill, Copernicus in his deepest soul was not a scientist at all; in every sense he was the product of the Middle Ages. His argument was still based on the perfection of God and the circle rather than on the testing of hypotheses by observation, and clearly he accepted the concept of ancient wisdom being embodied in unquestionable dogma: "We must agree to follow strictly the methods of the ancients and to keep to their observations, which we regard as a holy Testament. To those unwilling to trust the ancients implicitly, the doors of my science must be closed."

He wrote further that he recoiled in horror from any suggestion that the heavenly bodies might not move in perfect circles; such a concept was not up for discussion—the only point of the contention was *what* moved in *which* circles. His imagination was fired in the beginning by the possibility that putting the Earth in motion around the sun might lead to a simplification of the Ptolemaic system with all its circles upon circles; his imagination was mired in the end by the same old morass of faith and belief, of dogma and testament. For when he put the sun in the center and moved the earth about it in a perfect circle, the system he thus created did *not* provide agreement with the observations! He was immediately driven to create his own epicycles for the planets, and the supposed simplicity of the heliocentric system evaporated under the weight of observation and calculation. By this time, however, like so many of us today, he saw in his own system the workings of God, and died blind to its faults.

The real scientist at that time was Tycho Brahe, a Danish astronomer who believed that the Earth must be at rest in the center of the universe, not so much because of religious convictions as because of the lack of apparent motion of the stars (as determined by the trigonometry of Hipparchus and the best then-current astronomical observations). His own ideas actually incorporated the worst of both systems, with the Earth at the center and the planets revolving in epicycles around the sun, which in turn revolved around the earth, but the great difference between Brahe and Copernicus was that Tycho didn't worry so much about reconciling his ideas with those of the ancients and with the word of God—he concentrated instead on improving the observations and testing all theories against them. He spent his life devising and improving astronomical instrumentation, and using these instruments to measure star and planet locations with greatly improved accuracy. Toward the end of his life he hired Johannes Kepler, a German astrologer and metaphysicist skilled in mathematics, to provide the detailed calculations which would ultimately test the theories of the Greeks, of Copernicus, and of Brahe himself.

This concept was a wholly new one in the history of human thought. Even after the early Greeks had recognized the value of devising mathematical descriptions of real events, it did not necessarily follow that the mathematics must or even might describe precisely what was observed. For the universe was not thought of as a mechanical system, but as one imbued with a metaphysical life: angels or spirits moved the heavenly bodies, or how else could they continually move? And why should not the angels and spirits sometimes push a little harder and sometimes a little less hard? Why should they not sometimes vary their direction? Even those thinkers with increased sophistication kept subconsciously this atavistic precept, so that although they might not accept the notion of heavenly pushers it also never occurred to them that mathematical exactitude was to be expected. In this respect particularly, Brahe and Kepler brought us to a new level of understanding.

And in that competition which lasted many years of long days and nights poring over meticulously scribbled numbers and angles and times and dates, *none* of the three theories won. Kepler had come to Brahe as a confirmed Copernican, but the Copernican theory fit the data no better than did the other two. Indeed, as Brahe's observations became more accurate, more and more epicycles had to be added to the Copernican model to keep it in line; it was no longer the simple and direct vision of a perfect God.

Eventually, after long years of playing with the numbers and the theories, Kepler hit upon the solution: the planets moved not in perfect circles but in ellipses, with the sun not in the center but at one focus of the planetary ellipse.

My God! That result came as a shocker. Gone at one stroke was the perfection of the circle and of a universe with the earth at the center. The Copernican concept had already been labeled heretical, his book condemned as forbidden reading (and remained so well into the nineteenth century); now there seemed nothing left to cling to, if the etherial notion of mathematical reasoning was to take the place of solid Scripture as the font of truth. But what was there about mathematics that made it so holy? Why take the word of a German scribbler of arcane symbols and of a type of reasoning based on ancient Greek mathematics which one didn't understand, against the word of God as revealed to his priests combined with the reasoning of ancient Greek philosphers such as the wonderful Aristotle? There was and is nothing obviously superior in the Keplerian method.

But then came Galileo, and finally Newton, and the Greek/medieval geocentric structure collapsed under the weight of observation and law.

At the beginning of the seventeenth century the telescope was invented by Hans Lippersheim, a Dutch maker of lenses for eyeglasses, who made the accidental discovery that a pair of lenses held in series magnified objects at a distance. He cleverly fit a tube around them to hold them in place, and he had the toy of the century! The idea spread through Europe, but not until it reached Galileo in Pisa did anyone think to look up at the stars with it. That last sentence is certainly not true, but nobody looked at the stars carefully enough, methodically enough, seriously enough, until Galileo did. Looking at the myriad lights in the sky with even a good modern telescope doesn't reveal any new universes; certainly the first people to do so with the first telescopes saw nothing new. But when Galileo looked at the five planets he saw indeed a new universe.

He saw that the planets were different kinds of objects than the stars: the stars remained points of light in the telescope, but the planets became disks of definite size. He saw that Jupiter had its own moons orbiting around it: so not only was Earth no longer the only object to have a moon, but clearly it could no longer be said that every heavenly body revolved around the Earth. He saw the pockmarked face of the moon, and no longer could anyone say along with Aristotle that the heavenly bodies were perfect. Even the sun had spots!

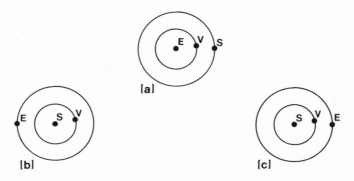

Figure 4.3.

(And this is an interesting phenomenon: sunspots are occasionally visible with the naked eye, and so must have been observed before the invention of the telescope; but they are nowhere mentioned in the ancient or medieval literature. The reason, I think, is twofold: in no cosmology or religion did anyone think of the heavens as being less than perfect, so the spots on the sun *ought* not to exist; and so, being seen, they were ignored. Second, they are not repeatable: they do not occur and reoccur regularly. So if indeed you happened to be the iconoclastic type and you did one day observe a spot on the sun and told people about it, the next day when they looked it wouldn't be there, so who would believe you?)

Finally Galileo carefully observed Venus throughout the year, and saw in his telescope that it showed phases like the moon. And this, it turns out, is impossible in a geocentric universe, as illustrated in figure 4.3.

Venus is seen in the sky only as a morning or evening star: it always lies close to the direction of the sun. Therefore in a geocentric universe it must always lie between the Earth and sun, as shown in figure 4.3(a). In this case the light we see it by, the light it reflects to us from the sun, is always glancing off it and so we should see only a crescent. In the heliocentric universe Venus and Earth may sometimes lie on opposite sides of the sun (figure 4.3b). In this configuration that light from the sun is reflected to us from the full surface area of Venus, and so we see it as a disk. If the planets lie on the same side of the sun (figure 4.3c) the reflected light will be the glancing rays that show us only a crescent.

Thus only the heliocentric universe can allow Venus to be seen both as a crescent and a full disk; only in this model are the phases possible. Galileo's observation of the phases proved the geocentric concept conclusively.

But only to those who were willing to look through the telescope and see what Galileo saw. And most people were not. "Most people," of course, didn't care. And even those people who did care about the structure of the universe cared more for what they had always been taught than for observational evidence, and so the Copernican doctrine (really the Keplerian doctrine) remained forbidden and blasphemous, and Galileo himself was finally arrested, tried, and convicted. To escape torture he confessed the error of his ways, recanted his arguments, professed the word of the faithful that Earth is of necessity and has always been at the center of the universe, stationary and immobile, a testament to God's favor.

The rumor persists that at his public recantation he muttered at the end, in a scarcely audible aside, "Nevertheless, it does move!" There is, I'm afraid, no evidence at all that he ever made such a statement. But, as Lady Brett once said, "Isn't it pretty to think so?"

SOURCES: Hoyle 1975; Hemingway 1926; Lovell, 1981

CHAPTER FIVE

•

THE AUTHORITY OF LAW

GALILEO SPENT the rest of his life a house prisoner of the Church. His spirit was broken, he was old and tired, he made no more trouble. But "forbidding" a book has never been the same as keeping it unread, and the ideas of Copernicus as modified by Kepler and proved by Galileo slowly at first, and then ever more rapidly, trickled, spilled, and finally swept over Europe in a thunderous wave. The proof of the heliocentric system was seen to be complete and incontrovertible, but the system itself was distinctly unsatisfying and indeed more than vaguely disturbing. It didn't, in fact, make sense.

In the Aristotelian scheme, the universe was a rational one. The heavenly bodies traveled forever in perfect circles because that is the nature of heavenly bodies, perfect and eternal. Here on earth, on the other hand, the natural motion of objects was seen to be distinctly different: they didn't move in circles but, if placed on a flat surface, stayed quietly still without any motion at all and, if left to themselves without any support, fell in a straight line down to earth—toward the center of the universe. Place a rock on a table and it stays there; raise it up over your head, let go of it, and it doesn't move around in circles; it just falls down.

That made sense according to Aristotle. The natural motion of heavenly and earthly objects could reasonably be expected to be different: the heavenly objects were ordained to move forever in perfect circles, while objects here on earth are intended to move toward the center, which on the surface of a spherical world defines the direction "down." No *reason* for this is necessary; that's simply the way the world is, and it's a perfectly reasonable way for the world to be.

But in a heliocentric universe all reason seemed to be thrown literally to the stars. If the earth is not the center of the universe but is instead a body wildly careening through space, why should objects tend to fall toward its center? What possible significance can the center of such an unstable object have? Objects might reasonably fall to the center of the sun, or perhaps imitate the earth in its elliptical (!) motion around that center, or even sail off to the stars—but they don't. They fall down. And how on earth was that possible, or reasonable, or understandable?

The intellectual community of Europe reluctantly accepted Galileo's proofs, they accepted the fact of a helicoentric universe, but they were not happy with it. It seemed less a triumph of reason and science than a retreat to the dark ages of mysticism and incomprehension. And then

> God said, "Let Newton be,"
> And all was light.
> (Alexander Pope)

Newton appeared to be a giant because (as he said) he stood on the shoulders of other men, such as René Descartes and Robert Hooke, who first proposed the laws we associate today with Newton's name. With these men we come at last to the concept of a universe governed by law rather than by the whims of the gods, a universe knowable and understandable through observation rather than through inspiration and dogma. Descartes stated this quite firmly when he said that he would "never accept anything as true unless I have conclusive evidence that it is so." He "took pleasure in mathematics because of its absolute certainty and the clear evidence of its reasoning." He determined "to believe nothing too firmly, and certainly nothing which has been passed on to me by example and custom."

He was a contemporary of Galileo, born in France, in Brittany, in 1596. He took advantage of the economic freedom his family's wealth provided him by leaving France for Holland, because he felt the intellectual climate there would be more receptive to the ideas that were beginning to grow in him; and so in due course the Dutch Protestants attempted to have him arrested and tortured to death, in an effort to save his soul and those of his disciples. It's hard to understand why his ideas should have engendered such hate, unless the reaction was based on fear in a time of change.

He (and Galileo independently) decided that Aristotle's idea of natural motion toward the center of the earth was not valid; instead they each came up with what we call Newton's first law of motion, the concept of inertia: an object at rest will tend to remain at rest and an object in motion will remain in motion along a straight line, unless a force is applied to it. (Newton graciously asserted that he stood on the shoulders of other men, but he preferred that these men should remain anonymous; he never acknowledged that Descartes and Galileo first enunciated this law, or that Robert Hooke first postulated the existence of gravity.)

Furthermore, both Descartes and Galileo stated that such laws of motion that might exist must be applied equally well to the heavens as to the earth—and this was a revolution of the deepest sort. It broke not only with the dogma of the separation of God and man (God created the heavens, on one hand, and the earth, on the other; He placed a firmament in the midst of the waters, and below the firmament is the place where man dwells while above it is the sphere of the aetherial bodies), it broke also with the evidence of the senses, with observation, for clearly the heavenly bodies are neither at rest nor in straight-line motion.

Descartes now went on to suggest the existence of another force which kept the heavenly planets moving around the sun. (It was agreed by this time that the stars didn't move at all, but only appeared to do so by virtue of being seen from a spinning earth.) Kepler a few decades before had visualized just such a force emanating from the sun, a system of solar rays which impinged upon the planets and pushed them on their way. But no mathematical analysis could provide agreement between such a postulated force and the observed planetary motions, and so Kepler had then tried to bring in the additional force of magnetism; William Gilbert had just published a book detailing the magnetic properties of the earth, so a generalization to all the planets didn't seem unreasonable. Still, however, the results of his calculations based on this theory couldn't quite be made to fit the observed facts.

Descartes now suggested instead that the planets were continually *falling* toward the sun, and that some linear motion originally given them at the time of their creation balanced this falling motion to result in their observed orbits. This was as far as he got; he was never able to visualize why the planets should fall toward the sun. The first person to suggest a reason was Borelli, an Italian doctor and mathematician, who took the giant step of suggesting a force emanating from the sun without the corporeal existence of Kepler's rays, a "pure" force which affected the planets without actually *touching* them in any material way. This concept of a "force acting through a distance" was seen again as a return to magic, it smacked of astrology and superstition. How can one body affect another if there is no connection between them? The idea seemed destined to die, and Borelli could suggest nothing to make it more palatable.

And then in 1666 Robert Hooke demonstrated the existence of just such a force. Hooke was born on the Isle of Wight in 1635, between Galileo's death and Newton's birth. He went to Oxford, worked for Robert Boyle, and there, sometime before he discovered Newton's Law of Gravitation, he discovered Boyle's Law relating the constancy of pressure and temperature in a gas, which became one of the foundations of atomic theory. He is the quintessential Unknown Soldier.

In 1666, in a public lecture to the British Royal Society, he took a pendulum at rest and gave it a push. Naturally it swung to and fro in the plane defined by its rest position, the point from which it hung, and the direction of the push. Nothing new in that. But then he pulled it back from its rest position, held it a moment, and gave it a *horizontal* push—and it moved instead in a circular orbit around its original rest position: it looked remarkably like a planet moving around the sun. Eventually, as it slowed, it returned to its central position of rest. Hooke suggested therefore that there must be a central force pulling it toward its rest position—since it ultimately ends there—but that in the meantime its motion is *around* the center. If it were not for the force of friction as the pendulum moves through the air, would it not continue to circle its center forever? Similarly, a central force emanating from the sun would constrain the planets to move in their ordained orbits.

Some ten years later he published this idea as a theory of universal gravitation, stating explicitly the remarkable idea that one and the same force governs the motion of the planets and the motion of objects here on earth. He went on to suggest that this force must be greater the closer the affected objects are, and that objects originally in linear motion will be bent by this gravity into a curvilinear motion "describing a circle, ellipse, or other more compounded (complex) curve," but he never formulated a mathematically exact description of gravity. He knew that it must vary with distance, but didn't arrive at the exact distance function and so was never able to correlate his gravity with the observed motions of the planets. This, finally, was Newton's achievement.

Isaac Newton was born the year after Galileo's death in 1641. By the time he grew up there was no longer any controversy about which view of the universe was correct: people were no longer burned at the stake for professing that the earth moves. The Church never recanted, never said it was wrong; it simply stopped talking about it and a new generation grew up wondering what all the fuss had once been about. (This replacement of one generation by another seems to be the necessary condition for what we call progress. Max Planck, in the twentieth century, told his contemporaries that his new and controversial quantum theory was waiting for this condition to be fulfilled—and he was right. Death, like taxes, has its uses.) And so Newton was free to wonder, not *if* the planets moved about the sun, but *how* they did so.

In 1665 the bubonic plague returned to England; it had been decimating the population sporadically for the past few centuries, disappearing and reappearing with neither warning nor apparent reason. In that year Newton graduated from Cambridge, and because of the plague he went not to London but back to his home town of Woolsthorpe, where he remained in intellectual isolation much as Einstein later did in the patent office at Berne. There is evidently something to be said for the practice, for in these two years he produced an edifice of physics and mathematics which remained unmatched until Einstein's work two hundred and fifty years later. He built this edifice on the foundation of Descartes' inertia and Hooke's gravity, although he never acknowledged either of them and, in fact, fought bitterly with Hooke over the ownership of gravity.

But even if we acknowledge the prior work of Descartes and Hooke, it is with Newton that all the bricks and stones are put together to form a cathedral of the intellect that dwarfs Notre Dame in the immensity of its conception and the eloquence of its statement. The universe was now, finally, a rational organization, subject to laws perhaps only dimly conceived but mathematically described, with results which could be quantitatively and accurately predicted. There was nothing in this mathematical construct which necessarily denied the Judaeo-Christian God, but no longer was it necessary to invoke this God in order to understand His universe—future generations were freed to decipher the origins and workings of a universe based on law, just as Newton himself had been freed (by Tycho Brahe, Johannes Kepler, and Galileo Galilei) to describe a universe not constrained to revolve around the earth.

•

CHAPTER SIX

THE ORIGIN OF THE SOLAR SYSTEM

T HERE WAS ANOTHER sense in which the revolution instigated by Copernicus and won by Newton freed future thinkers, allowing them for the first time to think rationally about the origin of the solar system, and that was the realization that the solar system in fact exists as a distinct entity in the universe. No longer was the universe a system composed of an earth-center surrounded by all the heavenly bodies; now it was known that our local system was the sun surrounded by the planets, and all the stars formed no part of this system. With this realization it became possible to separate the problem of the origin of the universe from the origin of the earth and its solar system, and thus to make a start on each of them.

And further, the concept of the solar system as a machine grinding away with its own inexorable pace from day to day under the driving force of gravity led one naturally to think of its origin, for consider this problem: Newton had said, and effectively proved, that gravity is a force between any two objects that attracts them directly toward each other. Then why does the earth not simply fall directly into the sun instead of circling around it forever? The answer is illustrated in figure 6.1.

Imagine a tower, and from this tower you drop a ball. It drops straight down to earth, of course. Now on top of the tower you have a musket, and you fire a musketball straight out horizontally. What happens to it? It falls to earth, but not *straight down:* it follows curve A in the figure. Now

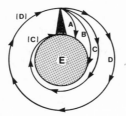

Figure 6.1.

put a stronger charge of powder in your musket and fire it again: the higher velocity of the musketball carries it in curve *B*. Theoretically, even in the seventeenth century, it was possible to keep increasing the powder charge and to imagine the increasing flight of the ball. Finally, in case *C*, the ball would fly completely around the earth before falling to the ground at the base of the tower. And so we try once more with an increased charge, and this time, case *D*, the musketball comes flying all the way around back to where it started. What makes it come back? If only the initial force of the powder charge were acting on the ball, it would have flown off in a straight line to infinity; it is the balance between this initial force and the continuing force of the earth's gravity that bends it around the earth. In fact, during its entire trajectory it is continually falling toward the earth, and case *D* is no different in this regard than the others. In each case the trajectory of the ball has been determined by two forces: the initial push (the musket powder charge), and the pull of gravity.

Now once the ball comes back to where it started, in case *D*, Newton's equations tell us that it comes back with the same velocity it acquired originally when fired from the musket (in the absence of any air to slow it down during its journey around the earth). It then starts its second journey with the same velocity and going in the same direction, and if these initial conditions are identical then the result must be identical: cause and effect clearly mandate that the ball will orbit once again around the earth and return to its starting point and will, in fact, continue to do so forever. This is what the earth does around the sun, and the moon around the earth: they are held in their orbits by a perfect balance between gravity and some initial push.

Once understood in this way, the next question is obvious: what provided the initial push? In other words, how did the solar system originate?

The first attempt to answer the question was by Immanuel Kant, the German philosopher, author of *The Critique of Pure Reason* and other works of light-hearted imaginative fantasy (as Tom Lehrer might say). In 1755 he published his *Universal Natural History and Theory of the Heavens* in which he pictured the solar system as condensing from an immense cloud of gas under the influence of gravity. He was no scientist, and his theory was so vague as to be untestable and therefore worthless in itself, but it was followed up by the French mathematician Pierre Laplace nearly half a century later and given form and some substance.

By this time William Herschel in England had discovered not only stars in the heavens, but nebulous, expanded luminous structures which he called nebulae. These we know today to be galaxies composed of up to a couple of hundred billion different stars, but in the eighteenth century they were thought to be gaseous structures about the size of our own solar system. Laplace, in fact, suggested that they were solar systems in the act of forming. He suggested that such a large cloud of gas would be in rotation, and this rotation would provide in essence the "initial push" necessary to balance gravity as the system condensed. For as it condensed, not only would the gravity increase—since it depends inversely on the square of the separation between each of the particles—but the speed of rotation would increase due to the conservation of angular momentum (as we discussed in chapter 2). The result would be an inequality of forces. A nonspinning cloud contracts under the sole influence of gravity, which is spherically isotropic, and so no direction is preferred: the result is a sphere. But if the cloud is spinning around the vertical axis, another force is produced: centrifugal force, directed radially outward along the line of the cloud's equator. This tends to diminish the inward pull of gravity in the plane extending out along the equator, with the result that the cloud contracts into a *flattened* or *nebular* disk around a central bulge: the central bulge, Laplace went on, would contract into a spinning sun while the nebular disk, as it contracted further, would shed concentric rings which would then individually condense into planets—each of them formed with the proper initial push (angular momentum) to exactly counterbalance the inward pull of the sun's gravity.

The hypothesis was tremendously popular and was accepted as at least a basic version of the truth well into the nineteenth century, but finally two quantitative factors destroyed it—one theoretical and one experimental. James Clerk Maxwell demonstrated theoretically that it would be extremely unlikely for such nebular gaseous rings to coalesce into planets; the most they might do is form small grains of dust like the rings of Saturn, since the total mass of the planets is insufficient (if it were once dispersed in a disk around the sun) to pull itself together gravitationally into planetary clumps. This was disturbing, but the final death knell was sounded by measurements of the sun's rotation. Quantitative calculations indicated that the sun, at the center of the contracting cloud, would be rotating the fastest; it would end up rotating aproximately twice a day. But measurements showed that it rotates only once every twenty-six days. And so the hypothesis died. But remember the Phoenix.

As numerous attempts to refute these objections foundered, it became clear that an entirely different type of origin was necessary. Early in the twentieth century T. C. Chamberlin and F. R. Moulton (who carried out the stringent calculations which showed that the sun's measured rotational speed was totally incompatible with a Laplacian origin) conceived of a truly new idea, an encounter between two stars. Actually, it was not a truly new idea; it seems as if there is almost never anything like a truly new idea in science. This is because ideas that are essentially different from anything that has come before are simply too much for one individual to grasp, refine, and mold into meaningful theories. Thus several people thought of the idea of gravity—an invisible and incomprehensible force, really, with nothing in the nature of matter to indicate the why and wherefore of its existence—before Newton was able to describe it properly and put it to use. And the stellar collision theory of Chamberlin and Moulton goes back to an old idea by a Frenchman named Buffon that the solar system was created by a collision between a comet and the sun. His idea was that a massive comet, instead of passing by the sun as comets are seen to do, by a cataclysmic misfortune actually fell into it. He thought the result would be a massive explosion which might spew solar material out into space where it could condense into the planets: some of the material would fall back into the sun, some would be lost to space, but some would have the proper balance of forces to fall into orbit.

This theory never became terribly popular, not so much because of its intrinsic faults as because of an antibiblical fervor that was sweeping through scientists. During the preceding centuries the biblical story of creation had dominated science in depicting astronomical theories, and had finally and with great effort been overthrown. In earth history, too, the Bible had been supreme: the geological story of our planet was thought to have been produced by the cataclysmic events such as Noah's Flood that the Bible recounts. But during the eighteenth and nineteenth century this philosophy of *catastrophism* gave way to that of *uniformitarianism* in geology, which stated that the history of the earth was to be understood not in terms of unique worldwide events but rather in terms of forces and processes that we see operating today—wind and tides, erosion and sedimentation. The idea was extremely fruitful, and with it our understanding of earth history for the first time began to flower; it was therefore recidivistic to think of an unlikely catastrophe as the start of the whole system.

Eventually harder evidence was brought to bear on Buffon's cometary idea, when comets were shown to be gigantic but incredibly massless objects, composed mostly of gas and dust: a comet falling into the sun would hardly raise a loud belch, let alone an eruption massive enough to form the planets. From that point all interest was concentrated on the Kant-Laplace idea of nebular condensation, and it wasn't until that idea was demolished by Chamberlin that he and Moulton thought to resurrect the Buffon idea, replacing the comet with an impacting star.

It was realized by this time that the sun is itself a star, and rather a typical one. The range of stellar sizes, however, is quite large: from much smaller than the sun to thousands of times larger. Furthermore, although they appear to be fixed in space, the stars are actually in motion relative to each other, a concept that depends on the antiquity of Creation—which is itself an interesting story of argument, error, experimental discovery, and the unshackling of the mind from ecclesiastical fantasies.

CHAPTER SEVEN
•
CREATION, ANTIQUITY, AND TIME

U NTIL WELL INTO the nineteenth century the general consensus was that the biblical estimates of the date of creation were at least approximately correct. Lightfoot's calculation which put the creation at precisely 9:00 a.m. on the morning of Sunday, October 26, 4004 B.C. might have been thought to be taking things a bit too literally, but most people felt that Bishop Usher's estimate in 1654 of simply the year 4004 B.C. (obtained by summing the ages of each generation of patriarchs listed in the Bible and then adding in the years of recorded history) was at the least a good order of magnitude guess. His chronology was actually inserted as "official" commentary in reference editions of the Authorized Version of the Bible. History, after all, faded away into oblivion at not much more than two thousand years, and if we had come from the barbarity of the ancients into civilization in two thousand years, another four thousand or so before that seemed a long enough time for the world to have existed.

The scientific conflict in this area actually began with the first serious geologist, an eighteenth-century Scottish gentleman farmer and physician named James Hutton. It was he who formulated the principle of uniformitarianism, when in 1785 (exactly two hundred years ago as I write) he stated that the earth we see today was formed by physical and geologic and chemical processes operating slowly over very long periods of time, and still operating today. For example, arguing for the principle of mountain erosion by running water, he argued that there can be seen "a chain of facts which clearly demonstrates that the material of the wasted mountains have traveled through the rivers," and that "there is not one step in all this progress that is not to be actually perceived" today. Such processes as the wearing down of mountain chains by continual river erosion must clearly take immensely long periods of time, beyond the limits of human imagination to visualize. Hutton made no estimate of the age over which these processes were occurring, saying simply that he found in the geologic record "no vestige of a beginning, no prospect of an end."

He didn't mean to imply by this that the earth actually existed throughout infinite time, but that the time scales were so long they were beyond human measure. Such an abstract statement did not reawaken any Church-science antagonism, there was no clerical reaction at all. The trouble began in the next century, with Darwin.

Charles Darwin's wild idea that species have changed and evolved into other species was not compatible with a few-thousand-year time scale, for we do not see such changes occurring today nor have we through all recorded history: cats were cats in the days of the Trojans, dogs were dogs, and people were people, not apes. It therefore became an inescapable requirement of the theory of evolution that the biblical time scale must be wrong; Darwin's own calculation, based on geological processes, of a 300-million-year age for the earth was attacked by several leading geologists as "geologically absurd" and "amateurish," and he retreated to the generic use of the term "millions of years," meaning simply a very long time, rather than insisting on a quantitative estimate. The importance of his arguments in this connection is that they stimulated people to think quantitatively and scientifically about the age of the earth.

Darwin published his *Origin of Species* in 1859; in that year William Thomson, professor of physics at the University of Glasgow, was thirty-five years old and was concerned with the question of the burning sun. At the time we knew nothing about the source of the sun's energy. It had been suggested that it burned with a fire of the type we find normal here on earth, simple chemical oxidation, but calculations indicated that the mass of the sun could burn for no longer than about ten thousand years by such a mechanism.

This was fine so long as the earth was thought to be only six thousand years old, and if the Redemption was due long before another four thousand years might pass, but it wouldn't do if the earth were millions of years old. A few years previously Thomson—who was to become one of the premier physicists and, after being raised to the peerage in 1866 as Lord Kelvin, the supreme (though often wrong) voice of scientific authority of the nineteenth century—had heard a British Association lecture by James Waterston in which the sun's energy was supposed to be derived from heat liberated by the infall of meteors. Kelvin (to use the name by which he is remembered) spent the next year calculating the quantitative effect of such a mechanism and convinced himself that it was a more reasonable energy source than chemical reactions could be. The observed flux of meteors with which the earth interacts was too small to provide the desired effect, but a postulated cloud of such meteors orbiting well within the earth's path was possible. This led to the concept of an aetherial vortex surrounding the sun, in which the meteors slowly and continually spiraled down into the solar atmosphere, where the friction of their passage supplied the required solar energy flux.

But the hypothesized vortex would do more than bring the meteors into the sun: it would bring them in with a rotational velocity component. He made the assumption that this, in fact, was what caused the sun to spin; and from this starting point he was able to calculate the mass of the meteoric infall necessary, given the hypothesized vortical motion, to bring the sun to its present state of spin. The parameters set by this calculation gave him directly one further result, the length of time it would have taken for this meteoric mass to fall in: in other words, the age of the sun. This calculation showed that the process could have generated the sun's observed energy outflow for just about 32,000 years, and that it couldn't last for more than about ten times that span in the future.

He proposed this idea in a lecture to the British Association in 1854, and there it rested until the occurrence of two great events: the publication of *Origin of Species* in 1859 and the breaking of Kelvin's leg in 1860. The *Origin* got him to thinking about the necessity of longer time spans, and the accident put him into bed for months; during this time he rethought his ideas on solar energy generation. In 1854 Heinrich von Helmholtz had made new calculations on the possible origin of the sun as a consequence of the nebular theory of Kant and LaPlace, replacing the vague wording of their ideas with specifics in which a meteoric infall to the sun (which

Kelvin had visualized as merely supplying heat to a sun already formed) became the basis for the creation of the sun. In this view the sun and other stars form by gravitational collapse of meteoric bodies, and as they form they release gravitational energy sufficient to heat themselves up to the point of giving off heat and light. Kelvin eventually became convinced that this idea was the correct one, and from the known mass of the sun and the measured intensity of heat and light radiated out per second, he calculated that such an energy source would last not 30,000 but 20 million years.

During the last half of the nineteenth century he also calculated the age of the earth from a totally different point of view. He argued that the solar system consists almost wholly of the sun iself, that the planets are in fact mere satellites of the central body and must have been created in some fashion either in conjunction with it or torn out of it in some manner; either way the earth must have begun its life as a molten ball of lava with temperatures greater than a thousand degrees centigrade. Obviously it is much cooler today, and therefore one might find it possible to calculate how long it took to cool down. He was helped in this by the observation of elevated temperatures in coal mines, from which he deduced that the interior of the earth is hotter than the crust. Using some simple measurements of the temperature differential as a function of depth and of the thermal conductivity of silicate rocks, he was able to set up the partial differential equations of terrestrial heat flow, and from these to calculate the rate at which heat was radiated away from the earth and lost to space. The result indicated that the earth was somewhere between 25 and 400 million years old. Further work by himself and others over the remainder of the century narrowed the error estimate and refined the final value to just about 25 million years, in excellent agreement with his estimate of the age of the sun.

Meanwhile, in the last years of the century, another approach was proposed by the Irish geologist John Joly, on the basis of the hydrologic cycle first proposed by Hutton. The question he raised was how the oceans became salty——or rather, he used the answer that had recently been arrived at by a variety of workers to delve into the question of time. The hydrologic cycle describes the continual passage of water through our environment: water evaporates from the oceans, forms into clouds, precipitates and rains back down onto earth again. The rain that falls on the oceans simply recirculates, but some rain falls on the continents. Here it seeps through the ground and collects eventually into streams, feeds into rivers, and ends up pouring once again into the oceans; but on this journey through and over the earth it dissolves a multitude of chemical species, from simple salts such as NaCl to complex minerals like $(Mg,Fe)_2 (Al, Fe)_5 Si_3 O_{10} (OH)_8$, so that the river water pouring back into the oceans carries with it a variety of dissolved ions. These ions are nonvolatile, and do not evaporate into the atmosphere; when the ocean water evaporates they are left behind. The consequence is that the oceans grow slowly richer in dissolved ions ("saltier") as time passes by.

Joly took the measured concentration of sodium ion in the ocean at the present time (1898), divided it by measured fluxes of that ion into the ocean via river water, and calculated that the process must have been going on for roughly 80–100 million years to bring the oceans up to their present salinity. The effort was a rough one, of course: he could not take into account possible variations in sodium flux due to varying erosion rates or ocean water surface area or decreasing sodium content of the land as it washed into the oceans, nor was the worldwide river flux known accurately, nor could he preclude a possible primordial salinity to the original oceans nor the possible transfer of sodium back to the land by such mechanisms as bubble formation and wind transport. But the method was based on known geological processes which could be seen to be operating today and which could be accurately measured, and as such it constituted a strong alternative to Kelvin's 20-million-year age scale.

Other geologists throughout the second half of the nineteenth century had attempted to establish time scales based on ongoing and observable geological processes, but their results had not nearly the same quantitative self-assurance. The best of these efforts is typified by the work of John G. Goodchild, curator of the Geological Survey at the Edinburgh Museum of Science and Art, who estimated the duration of the geological epochs by considering how long it took mountains to form, limestones to precipitate, and valleys to erode. He had also, however, to suggest long time spans for observationally thin geologic strata which showed varying fossil forms, and indeed to argue with very little evidence that the fossil-empty Precambrian must have extended over a "long" time in order to allow for the development (evolution) of the complex fossils which follow it. His conclusion that the earth was about 700 million years old lost its impact when he acknowledged that the uncertainties in his calculation were overwhelming and that, in fact, "all we can be sure of is that the records of the rocks fully justify us in claiming for the earth an antiquity so vast as to be far beyond the power of the human intellect to grasp."

Such qualitative arguments could not compete with Kelvin's precise analysis, but Joly's could and did. In 1900 he presented his arguments to the British Association. When Lord Kelvin replied in an address to the British Royal Society of London a few years later he was sure that he had answered with the full weight of the prestige of physics, and the stage seemed set for a continuing fight over the question of a "physical" twenty-five million or a "geological" hundred-million-year age for mother earth (while the last few biblical literalists continued to mutter in the wings). But then, out of nowhere, came Ernest Rutherford.

He had come from New Zealand to McGill University in Montreal, where he plunged into studies of the new science of radioactivity. In 1903 he (and independently Pierre Curie and Albert Laborde in Paris) discovered that radioactivity is accompanied by the emission of heat. Immediately there was a flood of letters to *Nature*, speculating on the possibility that this new discovery might vitiate Kelvin's limit to the age of a cooling earth and, at the same time, provide an alternative and long-lived source of energy for the sun. When Rutherford came to visit England in the spring of the following year, 1904, he was eagerly invited to speak at the Royal Institution about this new and exciting discovery. He was appalled to find Lord Kelvin in the audience, for in his speech he intended to present quite clearly his conclusion that there did exist within the earth sufficient quantities of radioactive elements for the heat they produced to completely overwhelm Kelvin's calculations. In estimating the cooling period of the earth, Kelvin had never thought to include any internal heat source; he had calculated simply how long a silicate body the mass of the earth would take to cool from an initial hot temperature with no subsequent heating. He can hardly be blamed for that, as radioactivity had been totally unknown and unthought of in the early 1860s when he had first published his theories, and in fact had been discovered only a few years before the 1904 meeting (and was still mostly thought of among his generation as "that Polish lady's Parisian plaything").

But Rutherford was now about to announce that the continuing disintegrations of radioactive elements in the earth would give off geologically

significant amounts of heat, therefore delaying the cooling time. His estimate suggested that the calculated age of the earth might be increased
from some twenty million years to as high as hundreds of millions or even
billions of years. If this were true it meant that Kelvin's estimate of the age
of the sun was also wrong, for an earth older than the sun didn't make
sense to anyone; and that meant, of course, that Kelvin's idea of the source
of the sun's energy must also be wrong, for he had shown that gravitational
energy could heat the sun for only twenty-five million years—it just
wouldn't last for billions of years.

Luckily, as Rutherford launched into his speech at the Royal Institution
that day, Kelvin fell asleep. Unluckily, just as he reached the point of heat
generation within the earth and the consequently great age of the earth,
Kelvin woke up:

"I saw the old bird sit up, open an eye and cock a baleful glance at
me! Then a sudden inspiration came, and I said Lord Kelvin had limited
the age of the earth (to 25 million years)—provided no new source of heat
was discovered! That prophetic utterance refers to what we are now
considering tonight (radioactivity). Behold! The old boy beamed upon
me."

So did the London newspapers. One daily, reporting the speech,
headlined "Doomsday Postponed!" Kelvin however, after his first beamish
smile upon this bumptious boy, thought better of letting him go untutored
to make his own way in this wicked world, and took him under his
avuncular wing. The two of them were invited to Terling, Lord Rayleigh's
home, for an Edwardian weekend; Rutherford wrote to his wife: "Lord
Kelvin talks (about radioactivity) most of the day, and I admire his confidence in talking about a subject of which he has taken the trouble to learn
so little. . . . He won't listen to my views. . . . but Strutt (the Fourth Baron
Rayleigh) gives him a year to change his mind. In fact they placed a bet to
that effect."

Strutt lost that bet. Although Kelvin began within the next year to
waver in a couple of recorded private conversations, he quickly recovered;
in 1906 he wrote a series of letters to the *Times,* and till the end of his life
he publicly sought to argue away the reality of nuclear energy. (In one of
his last papers, "An attempt to Explain the Radioactivity of Radium," published in the *Philosophical Magazine* in 1907, he tried to explain what he
viewed as the "monstrous" source of nuclear energy as simply energy

absorbed from the surrounding "ether"; he never again referred to the subject.)

Rutherford, meanwhile, working with the radiochemist Fred Soddy, extended his idea of the natural decay of radioactive isotopes, devising an experimental technique which enabled them actually to use it to measure the age of the earth. The technique, known as "radioactive dating," became one of the experimental cornerstones of modern geology. It's a perfect example of how it is possible to manipulate the physical world and tease out of it the Old Man's secrets, if only one is clever enough.

The decay of potassium forms the basis of one such method. One particular isotope of potassium, K-40, spontaneously disintegrates into the corresponding isotope of argon, Ar-40; the rate of disintegration is unalterable by time or external conditions: every 1.3 billion years (the "half-life" of K-40) half of the K-40 atoms will decay, and 11 percent of these disintegrations produce the atom Ar-40. The other 89 percent produce calcium-40, which is uninteresting; the Ar-40 product is interesting because of the different chemical natures of potassium and argon: potassium is a common rock-forming element, but argon is a gas. Keeping that in mind, consider the state of a molten magma under the crust of the earth, just before it erupts: the K-40 atoms naturally present in all earth materials will have been undergoing decay throughout the magma's life (indeed, since they were first created billions of years ago), and so there will be a certain number of argon atoms dissolved in the melt, kept prisoner there because the whole magma is locked under the crust of the earth, like a bubbling pot with the lid tightly clamped. But when the magma erupts onto the surface of the earth the effect is as if the lid has been suddenly lifted, and the magma bursts through at such high temperatures that the gaseous argon atoms bubble out and are lost into the atmosphere. This effect sets the geological clock ticking.

Because now the magma quickly cools down and solidifies into an igneous rock, with an initial content of generally a few percent potassium and zero argon. Obviously the ratio Ar-40/K-40 must equal zero at this point, which defines time zero for this particular rock. From this time on, however, any argon formed by potassium decay will be trapped within the crystal lattice structure of the rock, and so as the K-40 content continually decreases, the Ar-40 concurrently increases. In exactly 1.3 billion years half the K-40 will have decayed into Ar-40 and Ca-40, and so the ratio Ar-

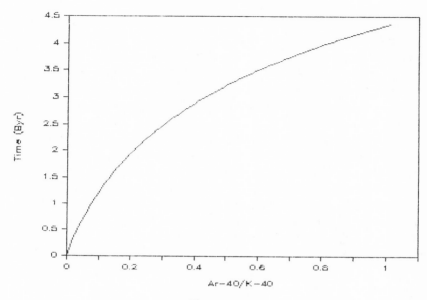

Figure 7.1.

40/K-40 = 0.11; as time goes on the ratio will continue to increase, as shown in figure 7.1.

The ratio Ar-40/K-40 is a single-valued function of the age of the rock: it is uniquely determined by the age and by no other variable, and so its measurement gives directly the age. The age of the earth, obviously, is at least as great as that of the oldest rock, and Rutherford and Soddy (using the decay of uranium instead of potassium) found rocks ranging in age from virtually zero up to about 2 billion years.

There are several different radioactive nuclides, each of which can zero in on the age of a rock from slightly different perspectives and different assumptions; among these are the decay of rubidium into strontium and of uranium and thorium into different isotopes of lead. These latter, although more complex since the daughter of the decay is present to some extent at time zero, give more precise results than does the potassium/argon method; with them the earth has been accurately dated at 4.6 billion years. This is an actual age rather than merely a limit, and is certainly accurate to within 0.1 billion years. The use of the uranium-lead system to obtain an actual age rather than a mere limit was first proposed by Fritz

Houtermans and (independently) John Holmes. The first accurate measure of the whole-earth age was made by Claire Patterson at CalTech in 1955, in which he measured the isotopic composition of lead in oceanic sediments—these being an average of the whole earth crust—and arrived at a value of 4.5 billion years with a possible error of 0.2 billion. Much work in the thirty years following has verified and improved the accuracy of his experiment.

The creation of the earth was thus moved, in the hundred and twenty years from 1785 to 1904, from nearly within the span of historical time and men's memory, first to an infinitude away, then back to millions and then forward again to hundreds of millions and finally to billions of years; from an imaginable to an unimaginably great number, from the recent past to the dim recesses of previously unrecorded time. The world—and the universe—had to be looked at now with different eyes.

SOURCES: Burchfield 1975; Faul 1978; Wilson 1983

CHAPTER EIGHT
•
STELLAR ENCOUNTERS

NEWTON'S CONCEPT OF GRAVITY was that of a universal force extending without limit throughout an infinite universe. The infinitude of the universe was in fact a consequence of that force, for when he tried to construct a model universe subject to gravity, he ran into a problem at its edges. Consider such a finite universe, in which every star is thought of as held in place by balanced gravitational forces, as in figure 8.1.

Figure 8.1.

The star system extends upward and downward and to the left, but star (Z) and its vertical neighbors mark the right-hand edge of this universe. Now conside stars (X) and (Y): they are pulled by the gravitational forces of their neighbors, but since the stars are evenly spaced and have equal masses, the pull of gravity in any one direction is balanced by an equal and opposite pull from the opposite direction. A universe constructed in this manner could be considered stable. It is, in fact, a necessary construction, given the seventeenth-century observation that the stars do not move relative to each other.

But look at star (Z). It is pulled gravitationally to the left by star (Y), and feels no balancing attraction to the right, since the universe ends to its right. It must necessarily then move to its left, eventually being pulled into collision with (Y), which itself would then feel no balancing force from its right and would begin to slide toward collision with (X), and in this way the universe must collapse upon itself and self-destruct. The only way to avoid this problem, Newton saw, was to envisage an infinite universe, one without such self-destructive edges.

But there was a problem even with this. Such a universe would find it difficult to maintain itself, since it would necessarily have negative feedback—any small distruption would tip the system into self-destruction. In figure 8.2(a)

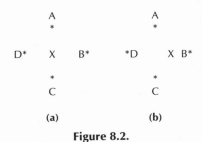

Figure 8.2.

the star X is held motionless by equally attractive forces from each of the stars ABCD, while in figure 18.2(b) a small disturbance has pushed X infinitesimally closer to star B. No matter how small this disturance, its effect is inevitably magnified since now the force XB is greater than its counterpart XD, and so X is pulled even closer to B, closing the logical circle which must eventually pull it directly into B. And B, in turn, would be disrupted from its own position of precarious balance and the same Domino Principle so beloved of political experts on Latin America would sweep through the heavens.

Newton was thus hung and left dangling by his thumbs through the bugaboo of science: inaccurate observations. This universe he constructed was the only one possible so long as gravity exists and stars do not move. He knew from Tycho Brahe's accurate measurements of planetary postions and his own mathematical reasoning that gravity exits; he concluded from similar measurements of stellar positions that the stars do not move. The former conclusion was correct, the latter was not.

Figure 8.3.

The error in the stellar measurements was extrinsic rather than intrinsic: the stellar positions were recorded with as much accuracy as were the planetary positions, but it was not realized how far away the stars are. They are indeed so very far away that the measurements were not able to discern changes in their position even when taken continuously over many years. By the end of the eigtheenth century, however, the first measurements of stellar parallax had indicated that even the closest stars were several light years away, with most stars lying hundreds and even thousands of light years distant. With this knowledge it was no longer necessary to postulate a universe with each star held motionless, glued in place by the perfect balance of gravitational forces. (Living in such a precariously balanced universe, it is no wonder that Newton and his contemporaries were profoundly religious men. Without a God to oversee such a mechanism, it couldn't possibly long exist.)

By the twentieth century we knew in addition, due to Rutherford's work, that the earth was billions of years old. Instead of a static universe thousands of years old, we had the possibility of stars in motion and periods of time amounting to billions of years to play with. In such a context, stars might even eventually collide, might they not? Although throughout our short history no such collisions have ever been observed, it is clear that the short duration of our recorded history and the vastness and emptiness of space combine to make the observation of such a stellar cataclysm extremely improbable. The probability of collision between two stars in random motion, ignoring their gravitational attraction for the moment, is a function of their relative distance and size, and can be roughly calculated by assuming one star to be at rest and the other in random motion. In figure 8.3 star A is at rest at a distance d from star X. The diameter of A is a, and we ignore the diameter of X. Then X, which is in random motion, will collide with A only if it moves within the angle ϕ (in two dimensions only). If the stars are 10 light years distant and the star A is approximately the size of the sun, the angle θ is on the order of 10^{-9}, and so the probability of collision is $10^{-9}/360$, or about 3×10^{-12}; extending the calculation through three dimensions drops the probability to about 10^{-20}. The effect of including gravity is to make the effective diameter of A larger, but not enough to raise the probability from the levels of the very small. And so we have never, in all our history, seen two stars collide.

Which is not to say that it can't happen, given the immensely long stretches of time available, but only that it is not a frequent occurrence. And so early this century Thomas Crowder Chamberlin and F. R. Moulton revived and modified the old Buffon idea about our solar system originating through a cometary impact on the sun. Although a comet doesn't have enough mass to orchestrate any solar-system-wide effect, obviously another star would. The original idea of an impact was further modified so that the rogue star made only a close encounter to the sun, its gravity raising immense solar tides and pulling them out from the sun as it passed, until a long filigree of solar material was stretched out into space and finally snapped. As the star passed away, some of this material might fall into it and be carried away, some might fall back into the sun, and some might coalesce into small solid bodies *(planetesimals)* which might fall into orbit around the sun and eventually coalesce into today's planets.

The basic impetus for this theory was the difficulty the Kant-Laplace nebular model had with the low rate of spin of the sun. In the Chamberlin-Moulton model the difficulty is, in essence, legislated out of existence: the formation of the sun is no longer coupled to that of the planets. One *starts* with a sun already formed, and if the sun formed without planets one could visualize it as forming nearly spherically symmetric with very low angular momentum to begin with. The high angular momentum of the planets—their high speeds of revolution around the sun—are provided by the motion of the passing star. The other Kant-Laplace problem, that of the difficulty of aggregating a nebula into solid planets, is just as great or even greater in this stellar encounter theory with its hot solar gases somehow condensing and aggregating into the planets.

The theory has been attacked on other grounds, however—mistakenly attacked. The argument is frequently proposed that such a stellar encounter is extremely improbable, and on this account the theory is at fault. Certainly such encounters are extremely improbable, but no matter how improbable they are one cannot argue against their singular occurrence. One can only argue that they cannot be usual or commonplace in our galaxy. If therefore we see that a significant fraction of the stars have planets circling them, we have to conclude that planetary formation is an integral and natural consequence of star formation, and the stellar encounter theory is dumped. On the other hand, if we see that our solar system is the only one in the galaxy, then it would be precisely a very improbable theory that we should be looking for. And at the present time, we just don't know. We are not be able to see earth-sized planets around even the closest stars, let alone those further away, with our present astronomical equipment. For all we know for sure, we may be alone.

This brings us to the next step in the development of a theory: we need some data. We've had enough hand-waving eloquence for a while; we have to step back for a moment from theoretical considerations and take a closer look at what the solar system actually looks like. What are the observed facts that any theory of its origin must explain?

CHAPTER NINE

•

THE SOLAR SYSTEM

BASICALLY, the solar system is the sun: looking at it from the outside, from anywhere else in the universe, that is all we would see. The sun comprises 99.9 percent of the mass of the whole system which orbits around it. Earth is one of the nine planets, forty-eight or more moons, thousands of asteroids, and innumerable comets which form this family; obviously it would be nonsensical to attempt to think about the origin of Earth as a problem separate from that of the whole system. The problem may be reduced, in rough outline at least, to one question: did Earth form as a natural part of the process of solar formation or as some later accident? And so it becomes necessary to know something about the sun.

All physical measurements in the solar system start with the *astronomical unit* (AU), the mean distance of Earth from the sun. Once we know this, we can calculate the size and mass of the sun and all the other planets. The first accurate method of measuring the solar distance was complex, but simple—simple because it involved only basic trigonometry, but complex because it had to be done on objects other than the sun and then recalculated for the sun. It can be illustrated schematically, as in figure 9.1, where two observers A and B at different points on Earth look at a distant object P.

Without going through the trigonometry, it can be seen that the angles *a* and *b* measured by A and B, together with their distance from each other, define the triangle ABP and therefore the distance to P. Unfortunately this method can't be applied directly to the sun because of the small angle *p* and the large size of the sun as seen from earth (the sun is not a geometric

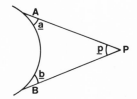

Figure 9.1.

point like P); but once the distance to any sun-orbiting object like another planet or asteroid is determined in this manner, the sun–earth distance follows from a direct proportion. Since the closer the object the larger the measured angles and thus the smaller the relative errors, the first measurements (in the seventeenth century) were made on Venus and Mars; in the early years of the current century an occasional asteroid such as Eros, which comes to within 16 million miles of the earth, was used. A different and more accurate method is used today: a radio pulse is aimed at Venus, and the time it takes to travel there, bounce off, and return can be precisely measured. Since radio waves travel at the speed of light the distance D is uniquely determined. The latest results give a mean solar distance of 149,598,000 km or 92,955,709 miles, in excellent agreement with the previous studies.

Once the distance is known, the size of the sun can be calculated from its apparent size as observed from Earth, and its mass from its gravitational effect on Earth. It turns out to be a sphere 1,392,000 km in diameter with a mass of 2×10^{33} grams. (Earth, for comparison, has a diameter of about 40,000 km and a mass of 6×10^{27} grams, or nearly 10^{22} tons.) The sun's size and temperature turn out to be rather typical in comparison with other stars—a statement which implies that stars can be organized and categorized by such measurements, and so they can. The two most basic astronomical observations are those of stellar colors and brightness, because of the ease with which they can be made and because they are intrinsically meaningful in terms of the physical condition of the stars. The color is a function of the surface temperature: cooler stars are just barely glowing red, while hotter stars burn with a yellow or white heat. The brightness of the star is a reflection of its size; obviously a bigger fire is brighter than a smaller one. When such observations are made on

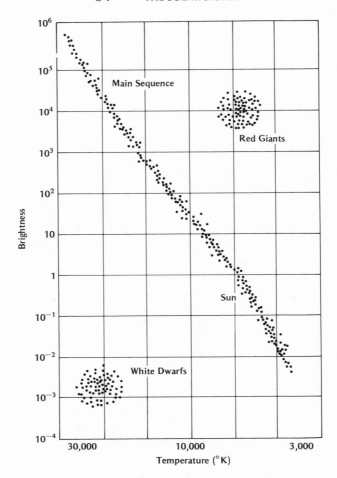

Figure 9.2. The Hertzsprung-Russell Diagram

a large number of stars, and corrected for distance effects, they can be plotted against each other as in figure 9.2.

Most stars lie on what is called the Main Sequence (the locus of "normal" hydrogen-fusing stars), and so does the sun. Neither its temperature nor its size is particularly extreme, and so clearly it is a rather typical star.

And clearly it is a very different sort of thing than the earth. Aside from its surface temperature of about 6500° centigrade and its massive size compared to Earth, the diameter and mass measurements show that it must be composed of quite different materials since its density (mass divided by volume) is only 1.42 compared to Earth's 5.52.

All the material of the universe is composed of ninety-two different elements (another dozen or so have been created by man, but they are radioactive and quickly disappear). Each element has its characteristic atom, with a definite number of protons and a varying number of neutrons in the nucleus, surrounded by concentric shells of electrons equal in number to the protons. The simplest atom, hydrogen, has just one proton, one electron, and one or zero neutrons; the heaviest naturally occurring atom, uranium, has 92 protons and electrons, and 146, 143, or 142 neutrons. These two extremeties of the atomic spectrum are both rare elements on Earth. The most common element on earth, oxygen, has 8 protons and electrons and usually another 8 neutrons; the other elements of which the earth is mostly composed, silicon, iron, calcium, aluminum, and magnesium, range from 12 to 28 protons and electrons. The sun, with a smaller density, must be composed of lighter atoms than these common terrestrial ones. But which ones? How can the material composition of a body nearly a hundred million miles away be measured?

In the nineteenth century it was found that when a gas is heated it gives off light at particular wavelengths which correspond to its atomic composition: each element gives a series of emission lines which serve as a unique fingerprint. This was not understood until the 1920s, when Niels Bohr—now generally revered as one of the most formidable quantum mechanics of our time, but then a fresh Ph.D. spending a year in Rutherford's laboratory in Manchester—suggested that every atomic electron was normally confined to a particular orbit around the nucleus, leaving it only when the atom was excited. At such a time the electron would absorb the excitation energy by jumping to a new orbit, making a "quantum jump," subsequently dropping back to its original position and emitting light with a wavelength corresponding to the difference between the two orbits.

Even without understanding the quantum mechanism of the emitted light, however, nineteenth-century workers were able to use the effect to determine the atomic composition of any glowing gas—and the sun itself is just such an object. The first measurements of the spectrum of sunlight showed it to be dominated by hydrogen wavelengths: 93.5 percent of the sun is hydrogen. The remainder was a puzzle. The wavelengths were unidentifiable with those of any element known on Earth, and in 1868 J. N. Lockyer and E. Frankland concluded that it was due to a new element which they named helium, after the Greek word for the sun. Quantitative measurements showed that helium accounts for 6.4 percent of the atoms in the sun, so that hydrogen and helium together form 99.9 percent. Clearly any theory of Earth formation is going to have to be more complicated than might have been thought at first: both the stellar encounter and nebular accretion theories would say, in their simplest forms, that the planets and the sun should have similar composition.

The composition of the sun, like its size and temperature, is typical of all stars, and can be understood in terms of the general evolution of the universe. Briefly, the universe we see was created in a Big Bang, when all material and energy catastrophically expanded from an initial state of infinite density. The expansion is still going on today, as we see by measuring the Red Shift of light from distant galaxies: the characteristic hydrogen wavelengths are shifted by the Doppler Effect, according to the velocity of recession of each galaxy. We also see a remnant of the primordial fireball in a low-level background radiation which pervades the universe, in perfect agreement with that predicted by the theory.

In such a sudden expansion, the most likely grouping for all the protons and electrons would be in the form of hydrogen, the simplest element. In addition, the initial high temperatures and pressures would induce nuclear reactions in which four hydrogen atoms would fuse together to form helium, and so the composition of the sun and stars is easily explained: as masses of the expanding cloud of gas become gravitationally unstable, they would condense to form stars made of hydrogen and helium, in just about the proportions they are seen. (References given at the end of this chapter discuss the Big Bang and resulting universe in considerably more detail.)

But further measurements lead to complications. Although 99.9 percent of the atoms in the sun are hydrogen or helium, there is still the remaining 0.1 percent to be accounted for; this consists of the heavier, more complicated elements like oxygen, iron, aluminum, and silicon, which could not have been formed in the Big Bang. Even worse, the planets are made of very little hydrogen and helium, and vary in their chemical constitutions among themselves. Let's take a look at them.

Compared to the electrons whirling around atomic nuclei, they're much better organized. They all move in very nearly the same plane, within just a few degrees of the plane of the sun's equator (except for Pluto, about which more later). They are arranged at distances from the sun which show a subtle and peculiar regularity, first noticed by Titius of Wittenberg, a contemporary of Kepler, and rediscovered and given a more precise form in 1772 by another German astronomer, Johann E. Bode, director of the Berlin Observatory, who emphasized its importance. The regularity is not immediately obvious, as is evident from a table of the planetary distances known to Bode, given in terms of the sun–earth distance, the AU.

Mean Solar-Planetary Distances

Planet	Astronomical Units (AU)
Mercury	0.39
Venus	0.72
Earth	1.00
Mars	1.52
Jupiter	5.20
Saturn	9.55

These numbers can be generated in the following way (there is no theoretical astronomical basis for this, it's a pure numerical manipulation):

Take a base number of 0.4 for the first number. Then add an increment of 0.3 for the second number. For the third number add the increment times two, for the fourth number add the increment times two squared, for the fifth number add the increment times two cubed, and so on. This results in:

First number		= 0.4
Second number = $0.4 + 0.3$		= 0.7
Third number = $0.4 + (0.3 \times 2)$		= 1.0
Fourth number = $0.4 + (0.3 \times 4)$		= 1.6
Fifth number = $0.4 + (0.3 \times 8)$		= 2.8
Sixth number = $0.4 + (0.3 \times 16)$		= 5.2
Seventh number = $0.4 + (0.3 \times 32)$		= 10.0
Eighth number = $0.4 + (0.3 \times 64)$		= 19.6

The argreement between the two lists of numbers is excellent, except that no planets were known to exist at a distance of 2.8 or 19.6 astronomical units from the sun. But a few years later, in 1781, a young English organist at the Octagon Chapel in Bath discovered a new planet. He was William Herschel, gentleman, physician, and amateur astronomer who named the new planet Giorgium Sidus in honor of King George III, and who was rewarded by a medal, a fellowship in the Royal Society, and an appointment as personal astronomer to the King—an appointment which induced him to quit his musical post before he discovered that his new royal position paid well in prestige but not enough in pounds sterling to live on. For a few years he managed to make and sell telescopes for a living, then luckily married a wealthy London widow and lived happily ever after.

When he discovered the planet Giorgium Sidus (incidentally, the name met resentment in international astronomical circles and was soon replaced with Bode's suggestion of Uranus) he at first thought it was a very small planet, for so it appeared to be as viewed from earth. The discovery was made by noticing that a faint star (the future planet Uranus) had moved its relative position with respect to the other fixed stars from one observation to the next. When he looked at it more closely, he could see that it widened in the telescope to a small disk rather than a point of light: clearly it was not a star. Continued observations indicated that it was in orbit around the sun, and therefore was a planet; but although it moved, it moved very slowly. Kepler's third law of planetary motion states that the square of the period of revolution of a planet is proportional to the cube of its distance from the sun, and so by measuring its period Herschel could calculate its distance. The result was 19.2 A.U.

The agreement between this number and that calculated by the Titius-Bode law for the furthest planet stimulated astronomers to wonder why there was just the one exception: an empty place at 2.8 AU. The Baron de Zach organized a society of twenty-four international astronomers specifically to search for the missing planet, and on the first night of the nineteenth century it was found—by Giuseppe Piazzi, director of the observatory at Palermo, who had not been invited to join the society. He noticed that night a very faint star in the constellation Taurus that he was sure had not been there before, and indeed on the following nights he observed it to move its position. He plotted it carefully for nearly six weeks, and then the winter weather closed in and he lost it. By the time the weather cleared again, he couldn't find it. Panic in Palermo!

During the succeeding months the brilliant mathematician Karl Friederich Gauss took up the problem of predicting the future planetary positions from a few observations, improved the methods in use, and the next year Heinrich Olbers of the Bremen observatory using his new method rediscovered Piazzi's planet. Its solar distance agreed exactly with the number 2.8 predicted by Titius-Bode, but it was very tiny, less than a fifth the size of the moon. Olbers continued to search the sky in its region, and three months later he discovered another small planet, and within a few years two more were discovered. Today we know of thousands of these *asteroids,* nearly all discovered by long-exposure photographs in which the stars appear as points but the moving asteroids leave short trails; though their individual orbits vary wildly, the mean solar distance of all of them is 2.8 AU, in perfect agreement with the Titius-Bode law.

So the Titius-Bode law, which no one understood in terms of an underlying astronomical or physical principle, was triumphant. And now things got even more exciting. Tracing the motion of Uranus around the sun, several observers saw that something was wrong; it was not moving in the perfect ellipse that it should. The first explanations were that since both Jupiter and Saturn lay between it and the sun—and they were known to be the largest of all the planets—the gravitational effect of these giants was influencing its circumsolar orbit. But by the middle of the nineteenth century the pertinent calculations had shown that although there was some such effect, it wasn't enough to account for all the observed deviation in Uranus' orbit. And simultaneously the word began to spread from all directions: there must be another planet out there, beyond Uranus.

Why not? There's no end to the Titius-Bode law: taking it one step further we have

$$0.4 + (0.3 \times 128) = 38.8$$

for the solar distance of the next-to-be-discovered planet, the one that was perturbing the orbit of Uranus. Two young mathematicians, U. J. J. LeVerrier in Paris and J. C. Adams in Cambridge began calculating the orbital parameters of such a hypothetical planet in order to pinpoint its location. Adams finished his work first, but had difficulty persuading the director of the Cambridge Observatory to look where his calculations pointed; up till this time, theory had always followed observation in astronomy rather than preceding it, and directors of things have a general reluctance to taking any first step. Finally the Astronomer Royal decided that the Northumberland telescope might be used. This was actually the largest telescope in England, and thus the best to use in a search for the feeble light of a faraway planet. In July 1846, the search commenced.

Meanwhile LeVerrier had finished his own calculations, was unable in his turn to convince any French astronomers, and wrote to J. G. Galle in Berlin. That same night Galle looked in the predicted place and found the planet, later named Neptune. (It turned out in retrospect that the Northumberland telescope observers had actually seen it first, but had failed to recognize it.) And nearly a hundred years later, in 1930, a Boston lawyer named Percival Lowell who had deserted his profession and city to build an astronomical observatory in Arizona in order to search for the elusive canals of Mars, duplicated the Neptune story by using observations of Neptune's orbit to find still another planet, Pluto.

But now the smooth sailing gets a bit choppy. Aside from the fact that Pluto's discovery was later shown to be accidental, in that its mass turned out to be too small to account for the variations in Neptune's orbit so that its resultant sighting was nothing less than fortuitous, observations of Neptune's period of revolution allowed its solar distance to be calculated, and the result turned out to be 30.1 AU instead of the predicted 38.8. Both Adams and LeVerrier had predicted Neptune's position in the sky on the basis of the observed deviations in Uranus' orbit and an *assumed* solar distance of 38.8; their assumption was wrong, yet their result was correct, illustrating an application in science of a truth basic to all fields of human endeavor: it's good to be good, but it's better to be lucky. The position predicted on the basis of a false assumption coincided with the true position only because of a fortuitous cancellation of various factors in the calculation, which happen to fall together periodically but not always: if Galle had looked for Neptune on another night, he might never have found it.

On the other hand, Pluto turns out to be 39.5 AU from the sun—in excellent agreement with the prediction for Neptune! Which leaves us rather on the horns of a dilemma: is the Titius-Bode law to be taken as an astronomical truth or a set of coincidences? If a truth, there must be an underlying principle, a reason for the planets to be assigned these particular orbits, a reason yet to be discovered and presumably of as much importance to astronomy as the quantum mechanical reasons for the precise electron orbits are to physics. But it might be only a set of coincidences, and then of course any effort devoted to a search for an underlying principle would be in vain, would be a lifetime of wasted research.

Can one decide which it is to be? Perhaps not, but let's summarize the case and see how it turns out.

Planet	Observed AU	Predicted (Titius-Bode)
Mercury	0.39	0.4
Venus	0.72	0.7
Earth	1.00	1.00
Mars	1.52	1.6
Asteroids	2.8	2.8
Jupiter	5.2	5.2
Saturn	9.6	10
Uranus	19.2	19.6
Neptune	30.1	—
Pluto	39.5	38.8

The excellent fit for the first three planets is not meaningful, because the law has been manipulated to fit them: a good fit can be manufactured to fit any three consecutively increasing numbers if there are no basic *a priori* principles to set restrictions to the imagination. If, for example, the distance to Mercury was 0.3 instead of 0.4 AU, we could write the law as: take 0.3 for the first distance, add 0.4 for the next, then again add 0.3 for the third planet, 0.4 for the fourth, and keep on in steady alternation between these two numbers. Such a law would be at least as simple and direct as the Titius-Bode law, and any reasonably increasing sequence of three numbers could be treated in a similar fashion, particularly when only "reasonable" agreement is insisted on. The law becomes impressive only with the fourth planet, for a given sequence of three defines the fourth and following numbers; it becomes rapidly more difficult to fit a higher series than three with a simple law. So the agreement for Mars, the asteroids, Jupiter, Saturn, Uranus, and Pluto is very impressive—unless one notes that we fit Pluto only by ignoring Neptune, and that Pluto itself was probably not formed in its present orbit (as will be discussed later).

What we have then is a remarkably precise fit of an empirical law to six experimentally determined points, with one gaping hole in the law and with one of the six probably not pertinent. If an underlying principle were to be discovered that would fix the two chosen parameters 0.4 and 0.3 independently, no one would argue with the reality of the law; we would instead be trying to understand the reason for the Neptune "hole" and for Pluto's addition at that precise orbit. But without such an *a priori* fix, the law seems more of a devil's temptation, a teaser to lead us up a blind alley where the only pastime is the manipulation of numbers for the amusement of the idle.

Not everyone agrees with this assessment, and I would be surprised but not mortified if some day the Titius-Bode law turns out to be the key to all creation. Most workers in the field, however, try to ignore it. There are more useful clues to the origin of the solar system in some harder planetary data (the asteroids will be considered separately).

Size, Mass, and Density of the Planets

Planet	Mass	Radius	Density	Moons
Mercury	0.056	0.38	0.98	0
Venus	0.81	0.95	0.93	0
Earth	1.00	1.00	1.00	1
Mars	0.11	0.53	0.72	2
Jupiter	318	11	0.23	16
Saturn	95	9.5	0.12	21
Uranus	15	3.7	0.29	15
Neptune	17	3.9	0.29	2
Pluto	—	0.25	—	1

All the data are given in terms of Earth values. The data for Pluto are very uncertain because it is so far from us, but it is certainly small, probably moon-sized. We'll have to ignore it for a while, then come back to it later. The masses of the first four planets are all roughly the same as Earth's, but smaller; the next four planets are all much larger. The radii follow the same pattern exactly. The densities of the first four planets are again similar to Earth's but slightly smaller, while those of the last four are suddenly *very* much smaller. The first four planets have from zero to two moons, the last four have many (Neptune is so far away that it is difficult to see its moons; it may well have more that are as yet undiscovered). So now we know that not only is the earth one of a system of planets, but that the system is of two distinct kinds of planets. The inner planets are known as the *terrestrial* planets because of their resemblance to Earth; the outer planets are often called the *giant* or *major* planets, for obvious reasons.

This difference in physical characteristics is reflected in their chemistry.

Chemistry of the Sun and Planets

Element	Sun	Terrestrial Planets	Major Planets
Hydrogen	93.5%	0.1	~93.5
Helium	6.4	5×10^{-10}	~6.4
Oxygen	~0.06	50	?
Silicon	~0.003	15	?
Iron	~0.002	20	?

The major planets are remarkably similar to the sun, composed mostly of hydrogen and helium—which fits well with our idea that the planets originated somehow in a tight relationship to the sun. But the terrestrial planets are quite dissimilar, composed of the heavier elements and very little hydrogen and helium.

To summarize, the solar system consists primarily of the sun, a large sphere of hydrogen and helium at high temperature, spinning slowly. Second, it is a system of planets orbiting the sun at distances large compared with their sizes. If, for example, the system is to be drawn to scale in a blackboard diagram and the sun is to be represented by a circle one inch in diameter, the closest planet, Mercury, would be a mere dot 4 feet away, Earth would be 10 feet away, Jupiter 50 feet away, and the furthest planet, Pluto, would be 400 feet away, further than the length of a football field. The planets all revolve around the sun in ellipses that are nearly perfect circles, except for Pluto, whose orbit is quite elongated. They all revolve in the same direction as the sun's spin, and very nearly in the plane of the sun's equator—except Pluto, which is tilted at an angle of 17°. They all spin on their axes in the direction of their rotation—except Venus, which has a small retrograde spin, and Uranus, which is flipped over on its side. Both these exceptions must be accounted for by postformational processes. The planets can be organized into two groups, the inner or terrestrial group and the outer or major group, each of which is characterized by similar mass, radius, density, chemistry, and number of moons—except for Pluto, which has unknown density and chemistry but which is certainly much smaller than the major planets (with whom it is placed by position).

In addition to the planets there are the asteroids, moons, and comets. The *asteroids* revolve around the sun in a swarm of planetary-type orbits, at roughly the distance predicted by the Titius-Bode law. They were at one time thought to represent a shattered terrestrial planet, since their individual densities are near the terrestrial value, but their total mass is nowhere near that expected; they are probably stuff left over from the formation process of the planets (to be discussed in detail later). The *moons* form miniature solar systems, revolving around their planets instead of around the sun, in the plane of their planet's equator—except for the outer moons of Jupiter, Saturn, and Neptune. They all spin in the same direction as their planet—except for some of the outer moons of Jupiter, one of Saturn, and one of Neptune, which rotate in the opposite direction.

The *comets* have not yet been discussed, and they're not going to be, for some very good reasons. We don't know much about them except that their orbits are very different from those of the planets and moons: they orbit the sun at all inclinations to the ecliptic and with no relation to circularity, receding so far away at their furthest distance that some of them might be considered not even bound to the solar system, approaching the distances of the stars. Some, in fact, travel in open orbits—they approach the sun only once and then whip away from its gravity and leave the solar system entirely. It's possible that at least some comets were never really parts of our system at all, travelers from the deepest reaches of interstellar space, ships that pass in the night. If so, they are not related to the solar system by birth and therefore aren't pertinent to arguments concerning its origin. Obviously there's a lot of uncertainty here, but that in itself is sufficient reason to say no more about comets.

For much the same two reasons we'll now dispense with further discussion of Pluto. We know so little about it, and it's likely that its peculiarities are unrelated to solar system formation. It may turn out to be not so much a rogue planet as an escaped and recaptured moon of one of the major planets, probably Neptune, whose orbit Pluto has approached quite closely in the past eons. I'll say a bit more about this later.

One final point needs mentioning. While the sun is spinning slowly, the planets are revolving around it rather rapidly. A way of putting this quantitatively is in terms of angular momentum, as pointed out by Chamberlin and Moulton: the sun, with 99.9 percent of the solar system's mass, has less than 3 percent of its angular momentum—more than 97 percent is carried by the planets.

These are roughly the facts that any model of solar system formation must account for. Now let's look at them in a little more detail and investigate some of the questions raised.

SOURCES: Fisher 1977; Weinberg 1977

CHAPTER TEN

•

THE SUN

WHEN LORD KELVIN suggested that the sun was some tens of millions of years old, and that its heat originated in the conversion of gravitational energy as it collapsed in upon itself, he was answering in one line of reasoning a host of questions that had never been answered before: how had the sun formed? how old was it? what was the source of its energy?—and he was very nearly right.

The energy source is the key to the sun's mystery. Kelvin's suggestion was the first reasonable explanation of a means to provide that incredible outpouring of light and heat at a steady rate for millions of years, upon which the continued existence of a habitable earth depends. His theory failed only when Rutherford showed that the earth, and therefore the sun, was actually *billions* of years old. The true answer involves a different kind of energy entirely, the nuclear fusion of hydrogen into helium and the $E = mc^2$ concept of Einstein. And it begins in the empty universe of interstellar space, where stars are born.

But nothing is truly empty. In my own research with primordial rare gases in meteorites and the Earth I spend a lot of time trying to achieve a good vacuum in a mass spectrometer, pumping the system down to pressures about one-billionth of the atmosphere. In one cubic centimeter of such an excellent vacuum there are still more than ten billion atoms floating around; "empty" interstellar space is much emptier than this, but even so there are always a few atoms per cubic centimeter, and most interstellar material exists in the form of "clouds" with densities of perhaps 50 atoms/cc. The temperatures are very low, on the order of 200° below zero centigrade, about 75° on the absolute (or Kelvin) scale, where zero is defined as the (hypothetical) temperature at which all atomic motion ceases, and so there isn't much random atomic motion to oppose the mutual gravitational attraction of the atomic particles in the cloud. But the cloud itself is so small that it still hasn't sufficient gravity to pull itself together. Instead it drifts through the galaxy, meeting other similar clouds, and as these clouds merge, coalesce, and grow, they reach finally a mass sufficient to begin gravitational collapse. The actual triggering mechanism for the collapse is still a subject for argument; we'll discuss it later.

Once the collapse begins, by whatever means, it necessarily accelerates unstoppably. A gravitational force is operating between every two particles in the cloud, directly proportional to their masses and inversely proportional to their distances. Once collapse begins the distances necessarily shorten and so the force must increase, the collapse accelerates, the force continues to increase, and so this increase of appetite grows by what it feeds on. As it collapses the gravitational energy is converted into heat, but this is quickly radiated away by the tenuous gas and the temperature remains low.

After a few million years the cloud has compressed to a point where it can no longer effectively radiate away the heat being generated by its collapse; the cloud is effectively opaque, trapping its own heat, and so the temperature rises. As it rises, so also does the gas pressure, and the collapse is slowed. The cloud inevitably arrives at a stage where it is hot enough to glow like a star. This is actually the process envisioned by Kelvin, but it isn't the end of the story. The collapse has slowed by this time but hasn't stopped, and the continued slow squeezing of the protosun, for that is what it now is, releases more gravitational energy than can be radiated away, and so it continues to heat up.

As it does the atoms within the cloud are being squeezed together with more and more energy, their velocities reflecting the increasing temperature, and when that temperature reaches a few million degrees they are hitting each other hard enough to break through the repulsive barrier of their positive charges and fuse together. Two hydrogen nuclei (the electrons have long since been stripped away by the heat) collide and rearrange themselves into a heavier isotope of hydrogen, deuterium; this in turn fuses with another hydrogen to form a different element, helium-3, and one more fusion turns it into helium-4. Here the process stops, for helium-4 is a particularly stable atomic nucleus, but it has gone far enough. The four original hydrogen nuclei weigh slightly more than the resulting helium-4 nucleus, and the mass that has thereby vanished reappears as energy according to $E = mc^2$, energy sufficient to fuel the sun's continual outpouring of light and heat for billions of years.

This idea of a nuclear fuel for the star's furnace was arrived at independently by several workers in the 1930s. Hans Bethe was probably the first to quantify it accurately, but as far as I know the first person to think of the concept was Fritz Houtermans, a young physicist then beginning his career as assistant professor at the Berliner Technische Hochschule, who later was to gain the distinction of being the only scientist in the world to be arrested on spy charges, tortured, and eventually released by both the Gestapo and the KGB. (He had fled Germany just two steps ahead of the Gestapo in the late '30s, when Rutherford obtained for him a job in Russia. But the KGB, germanophobically suspicious of all such refugees at that time, arrested him. When Germany and Russia became allies in 1939, the KGB as a gesture of secular germanophilic patriotism turned him over to the Gestapo. He somehow survived all this, and after the war founded the excellent Kosmophysik establishment in Berne.)

He told me once that he was intoxicated at the time (the early 1930s) with the fermenting, bubbling-over ideas and excitement of nuclear reactions and radioactivity, and after weeks of concentrated study he finally took an evening off to relax. He was lying on his back on a Bavarian mountainside, his arms around a girl; she was cradled against his chest, and they were both looking up at the stars above, thinking their thoughts, when he suddenly began to breathe deeply and tremble with excitement. She lifted her head, looked into his eyes—and he pushed her away, sat up, and shouted the German equivalent of, "Goddam it, I know what makes that sonofabitch shine!" It all came to him in a flash, and he then tried to explain it to her, he told me, but she somehow seemed put-out, she wasn't interested. He never did understand women.

This nuclear fusion process begins in the center of the protosun, where the temperatures and pressures are highest, and with its onset a new stage begins. The outflow of fusion energy from the core balances the inward-pulling gravitational energy, and the true sun, for now such it is, settles down into a controlled, stable system. The rate at which this nuclear fire will burn is controlled by the mass of the star; our sun has been in this

Photo 10.1 Margaret and Geoffrey Burbidge, William Fowler, and Fred Hoyle.

stage for four and a half billion years and will continue for another few billion before its engine runs down.

But this can't be the whole story. This picture gives us a stable sun, billions of years old, but a sun composed of pure hydrogen and helium. If we were interested only in stellar systems, we would undoubtedly be satisfied with a theory that predicts a sun to be 100 percent hydrogen and helium, compared to observational evidence that it is in fact composed of 99.9 percent of these two elements. But such a sun can have no planets composed of iron, aluminum, oxygen, and silicon. Where did these other elements come from?

The answer to this question came from the study begun in the 1950s between Willy Fowler, the nuclear physicist from CalTech, the English astrophysicist Fred Hoyle, and the Burbidges, Geoffrey and Margaret (photo 10.1) who together worked out the beautifully simple theory of stellar nucleosynthesis—the creation of the elements inside stars. In this theory past and future merge, for the sun was born of other stars that came before it, and will end as they did. Let's follow its future course.

The hydrogen fusion system at its center will in a few billion years give out, as it clogs itself with helium. In much the same way as ashes build up to suffocate a wood fire, the accumulating helium will interfere with the fusion process, the outpouring of nuclear energy will slow, and the balance between it and the sun's gravity will be disturbed. Once again the sun will begin to contract, and as it does the center will become even hotter. At a temperature of about a hundred million degrees the helium will begin to react, three nuclei fusing together to form the new element carbon. And once again the product, carbon-12, has less mass than the reactants, the three helium-4s, and so once again there is a release of nuclear energy. The sun by this time has expanded, its outer envelope pushed away by the increased heat in its center until it stretches out to a distance greater than 1 AU, swallowing up Mercury, Venus, and Earth. Although the center is so hot, the outer skin is now so far away that it becomes cooler than before and therefore more red than yellow, and our Main Sequence sun has become a red giant (see the Hertzsprung-Russell diagram, figure 9.2).

A periodic series of such processes continue to occur, with succeeding generations of fusion reactions producing heavier and heavier atoms—including oxygen, silicon, and magnesium. At the same time other reactions involving the addition of neutrons and radioactive decay produce most of the 92 elements. Eventually all the available nuclear fuel is used, the outer envelope is blown away in a last gasp, and the sun settles down to cool off and die.

In general terms, all stars follow this pattern. Stars much more massive than the sun burn their fuel more quickly, burning brighter and dying younger, and ending in a more massive explosion called a supernova. The end result of all this stellar activity is that during the star's lifetime some of the basic element hydrogen is converted successively to helium, to carbon, and to the other heavier elements; and that at the star's death these are thrown out into the countryside. Gradually, then, the universe which began as only hydrogen and helium is enriched in the heavier elements out of which planets—and life—may be formed. The sun itself cannot be one of the early stars of the universe, for then it would be only hydrogen and helium. Many billions of years must have gone by, during which countless stars processed their hydrogen and helium into the heavy elements, before ever the interstellar cloud which was to become our sun began to coalesce, collapse, and resolve itself into a roiling protosolar dew.

SOURCES: B²FH 1957; Wood 1979

CHAPTER ELEVEN
•
THE TERRESTRIAL PLANETS

MERCURY IS the closest planet to the sun, coming to within 30 million miles of it at perihelion (distance of closest approach). This is so close that the beautifully simple Newtonian equation for gravity breaks down and no longer accurately describes its orbit. All objects moving under the influence of a centrally directed force varying with the square of the distance must, according to classical Newtonian mechanics, orbit in an ellipse whose path does not vary with time. This is the path worked out by Kepler for all the planets and explained by Newton on the basis of a centrally directed gravitational force proportional to M x m/d². It was later discovered, however, that the elliptical path of Mercury *precesses:* the planet does not come back to quite the same spot every year, and so its continued path is not quite the same every year, as shown in exaggerated form in figure 11.1.

The actual precession is nowhere near as large as shown, the position of the perihelion moving by only 43 seconds of arc per century—a very

Figure 11.1.

small effect, yet one definitely not predicted by Newton. And yet Newton's physics was so all-encompassing up till the nineteenth century that it was nearly a religion, and astronomers in particular were so sure it must be correct that they interpreted Mercury's anomalous motion as due to the Newtonian pull of another (unknown) planet's gravity. The French astronomer LeVerrier, who in chapter 9 calculated the existence of Neptune from its effect on Uranus, now calculated where this Mercury-perturbing planet must be in order to account for the observed effects, and so sure was he of its existence that he named it before anyone found it. But this innermost planet, Vulcan, never was found; it doesn't exist. Instead, the discrepancy in Mercury's motion was explained by Einstein's general theory of relativity, which replaces Newton's force of gravity with curvature in space-time. The explanation of Mercury's orbital motion was the first great success of general relativity.

From Mercury's position so close to the sun, the size of the sun must appear enormous. It burns in the sky nearly ten times bigger than it looks from earth, searing the planet with incredible heat. Nearly all the planet's atmosphere has long since evaporated into space (Mariner 10 found less than a trillionth of an earth atmosphere there, composed chiefly of argon, neon, and helium), and at noontime the temperatures soar to over 500°C and then drop more than 700° at night—the most extreme range of temperatures to be found anywhere in the solar system. In daylight metals such as lead or zinc, if they were present, would melt in the heat, while at night any gases like carbon dioxide would freeze solid.

It's the smallest terrestrial planet; in fact, several moons in the solar system are bigger. Because of this small size and its position in space as seen from Earth—always close to the sun, appearing within an hour and a half of sunrise or sunset, a consequence of its orbit lying inside Earth's—it's extremely difficult to see clearly from Earth. It wasn't until 1965, when we were able to bounce radar waves off its surface and get a clear echo, that we knew anything about its rotation. In 1889 Giovanni Schiaperelli, director of the Brera Observatory in Milan, concluded from his observations that Mercury always showed the same face to the sun, so that one half the planet had permanent day and the other permanent night; but the radar echoes showed clearly a slow nonsynchronous rotation, one day lasting nearly 59 earth days, while it circles the sun once every 88 earth days. This slow rotation, compared to its period, leads to exotic effects. To

Photo 11.1. Taken in the late 1960s, this represents the best quality of Mercury visible from earth. NEW MEXICO STATE UNIVERSITY OBSERVATORY

an inhabitant of Mercury (all of whom comprise an empty set, because of the extremes of temperature) the Mercurian day would be longer than its year: if he measured the day from one dawn to the next, by the time the planet had completed one spin it would have moved two-thirds of the way around its orbit, so that the sun would no longer be where it was at the first dawn; the second dawn would be delayed until two complete orbital revolutions had been completed. In that time Mercury would actually have spun on its axis three times before the sun rose again. In addition, because of the two-year length of its day and the ellipticity of its orbit, the sun would not move uniformly across the sky, but would in fact reverse itself briefly before finally setting.

The difficulty of deciding anything at all about Mercury from Earth can be seen from photo 11.1, which shows a typical example of the clearest views of Mercury ever obtained from earth. Until 1974, in fact, when the Mariner 10 spacecraft slipped into Venus' gravity and used it to boost itself onward to Mercury, we knew virtually nothing about the planet. Then, in March 1974, Mariner 10 approached within 500 miles of Mercury, and showed us pictures like that shown in photo 11.2.

Mercury looks a lot like our moon, barren and pockmarked by the

Photo 11.2. This photomosaic of Mercury is from pictures taken by *Mariner 10* **in 1974.** JPL/NASA

same rain of projectiles early in the solar system's history. But unlike the moon, it presented the Mariner's instruments with evidence of a magnetic field and an iron core, showing that it had melted like the Earth and had chemically differentiated, with the heavy elements like iron sinking to the center. Its core is, in fact, the largest for its size of any object in the solar system; the planet can really be considered an iron sphere with a thin veneer of silicate crust.

Venus, although much closer to us, was just as much of a mystery until we went there: it is cloaked in thick clouds that never part to reveal the planet itself. Because of these clouds, which reflect the sunlight brilliantly, and because of its closeness to the sun, Venus is the brightest object in the sky (except for the sun and moon), ten times as bright as the brightest star. Like Mercury, Venus' orbit is inside Earth's so it is always seen as a solar companion, appearing within three hours of sunset or sunrise.

Photo 11.3. This picture of Venus was also taken by *Mariner 10* as it sped by on its way to Mercury. JPL/NASA

During the centuries following Galileo's first telescopic observation, Venus became recognized as Earth's twin sister. It has just about the same mass, size, and density. The thick clouds are evidence of an atmosphere (photo 11.3) and people began to speculate seriously about the possibility of life and civilization on the planet. But Venus turned out to be even more inhospitable than Mercury.

In the early 1960s, the golden age of space exploration, before we dissipated our national wealth in the jungles of Vietnam and New York, Mariner 1 and Venera 1 (from the USA and the USSR respectively) raced to be the first to Venus, and both vanished, lost in space because of technical malfunctions. In 1962 Mariner 2 achieved a successful flyby, and indirectly measured temperatures of at least 400°C at the surface, confirming what some scientists had long suspected. The Venusian atmosphere (a hundred times denser than earth's) is 97 percent carbon dioxide, shrouding the planet in a gas that provides a perfect greenhouse effect. The gas, like

Photo 11.4. Earth rising over the moon, as seen by *Apollo 11*. NASA

the windows of a greenhouse, is transparent to electromagnetic radiation with wavelengths characteristic of visible light—which is how the sun puts out most of its energy. So the sunlight goes through the clouds and heats the surface. The surface then reemits the heat with wavelengths in the infrared region, and the carbon dioxide is opaque to such wavelengths, so the heat can't get out. Trapped under the thick carbon dioxide clouds, the heat reverberates and the planet simply gets hotter and hotter.

This scenario is a warning to us on Earth: our atmosphere contains only about 0.03% carbon dioxide, but it is increasing rapidly because of our burning of fossil fuels and decimation of forest areas. In fact, the total carbon dioxide on Earth approximates the Venusian amounts, but here most of it is locked up in carbonate rocks and biogenic compounds such as oil and coal. During our history plant life has removed most of it from the atmosphere via photosynthesis, replacing it with oxygen. But as we burn our organic remains and kill off our plant life we are beginning to reverse the process, with possibly disastrous greenhouse effects in store.

It was again in the early 1960s, when Soviet and American radar

Photo 11.5. Mars, as seen in the best Earth-based photograph. The polar icecap and the dark *maria* are visible. R.B. LEIGHTON

observatories managed to bounce radio waves off the Venusian surface and obtain echoes, that we were able to measure the planet's rotation—and found that the damn thing spins backward, a fact which would destroy every conceivable theory of solar system formation if we allowed it to. We can think of scenarios in which the planets are formed by a variety of processes acting at random, creating planets which spin in all possible directions, and we can think of other scenarios in which the planets are created by one general process so that they spin with a greater degree of order, but for them all to spin in one direction while Venus alone spins in the opposite direction is just too much to explain. In the circumstances—which are that it simply *does* spin oppositely from all the other planets—we, like our ancestors, resort to faith: we suppose that its retrograde spin is not an original sin but was imposed upon it by later circumstances. In fact several ingenious theories have been proposed to explain how such an anomalous spin might have occurred, which we will deal with when we must.

In the later 1960s and early 1970s, *Mariner* 5 and *Venera* 4, 5, 6, 7, and 8 reached the planet, the *Venera* craft actually crashing through the atmosphere to land on the surface. They brought back a picture of an atmosphere 40 miles thick, with temperatures ranging up to 475°C. The atmospheric pressure is nearly a hundred times that of Earth's, similar to the pressure exerted half a mile deep in Earth's oceans. (For comparison, our World War II submarines could only go a few hundred feet deep before the pressure would crush their steel skins like eggshells.)

Later in the 1970s *Mariner 10* flew past the planet on its way to Mercury, and *Venera 9* and *10* achieved a landing on the surface soft enough to allow the cameras to function. The Mariner photos showed that the thick atmosphere is in constant hurricane condition with average wind speeds in the stratosphere over 200 miles an hour (on earth a hurricane is defined as a storm with winds of 65 mph or greater, while the really destructive ones reach wind speeds of usually under 150 mph). But when *Venera 9* and *10* landed, they found a relatively calm surface, the winds dropping off near the ground to only a few miles per hour. Shortly after sending back a few photographs which show a silicate surface, the equipment failed.

No wonder. Preliminary chemical determinations show that, aside from the carbon dioxide which forms most of the atmosphere, the clouds are composed of such stuff as sulfuric and hydrochloric acids and hydrogen fluoride. Because of this corrosive atmosphere, together with the high temperatures, it has not and never will be easy to gather much information.

The third planet, Earth, looks very much like Venus as it is approached (photo 11.4). The clouds swirling over its surface, however, do not form an impenetrable layer but break frequently to allow the sunlight (half as bright as Venus') to reach the surface of the planet. One might expect this to make the surface even hotter than Venus, but the absence of carbon dioxide from the atmosphere allows the surface heat to be radiated out to space and the planet stays at an average temperature of about 20 degrees C instead of over 400 degrees.

The chemistry of the planet is very similar to that of Mercury and Venus, with a silicate mantle and crust overlying an iron core, but the atmosphere is totally different. The major components of the atmospheres are:

Atmospheres of Venus and Earth

Planet	Carbon Dioxide	Nitrogen	Oxygen
Mercury	—	—	—
Venus	95%	1%	0.1%
Earth	0.03%	78%	21%

This tremendous difference in composition can be understood simply by looking at the surface of Earth, which is covered with water. This—and not its imaginary position at the center of the universe, as supposed by the ancients—is what makes the planet unique. Liquid water allows the development of life, and the totally different geological development of Earth, Mercury, and Venus is due in overwhelming measure to the presence of life on the first. The carbon dioxide that was lost from Mercury and that forms most of Venus' atmosphere is also present on Earth, but here it is chemically locked up in solid form as carbonate shells, skeletons, and rocks, formed and metamorphosed over hundreds of millions of years by the incessant hunger of life and the implacable burial processes of death. The floors of the oceans are covered with miles-thick sediments of carbonate microfossils and the sedimentary strata of our continents are layered with the metamorphic rocks formed from such fossils over the aeons. With all the carbon dioxide locked up inside the solid Earth, no significant greenhouse effect ever developed and so the Earth stayed cool enough for liquid water to remain on the surface; on Venus any water would all have evaporated, and would probably be photodissociated at the top of the atmosphere and lost to space.

The next and last terrestrial planet, Mars (photo 11.5), is again chemically similar to Earth and Venus but dry and dead. Not totally dry, and perhaps not totally dead: when the first *Mariner* flybys reached close to the planet we received photos that fairly clearly indicate the presence of flowing water there in the geologic past, valleys remarkably similar to the water-carved features we see on Earth. But when *Vikings* 1 and 2 landed on the surface of Mars in the summer of 1976, they found no trace of liquid water and only a problematical trace of life. The Martian landers carried a series of three sophisticated experiments designed to search the planet for signs of life, and the results have been interpreted by nearly everyone to indicate that the planet is dead. That word "nearly" remains, however, to give hope or irritation depending on one's preferences. Two of the experiments, designed to give an outpouring of gaseous by-products when a nutrient solution was fed to possible Martian creatures living in the dirt, did in fact do so, and have been interpreted by some workers as evidence of digestion of the broth by something actually alive in the dirt. Most of the scientists involved, however, interpret the reaction as a superoxidation process, the dryness of the planet arguing both against the viability of life and for the presence of more highly oxidising chemical forms than are stable on Earth. (The flux of solar ultraviolet radiation, which destroys organic molecules and which is not absorbed on Mars as it is on Earth by our atmosphere, also doesn't help.) The argument won't be definitely settled until we go back and finish the experiment, but geologically speaking the planet is certainly dead; living creatures never existed there in sufficient numbers to transform it as they have earth.

Surprisingly, early astronomers thought differently. Once the Church's proscription against thinking of the Earth as anything but the unique home of God's creatures became old fashioned, the pendulum swung in the opposite direction (as it always does: we seem to have a built-in antipathy to the happy medium—a tribute, perhaps, to our youthful optimism). William Herschel, the discoverer of Georgius Sidum, observing Mars and noticing a few swatches of white gossamer drifting across its surface, deduced correctly that he was seeing Martian clouds; and clouds, he thought incorrectly, mean rain and rain means water and water means life, and so he reported to the British Society quite casually that the Martian inhabitants of that planet probably enjoy an atmosphere similar to that here on Earth. He never knew that the clouds he saw were frozen carbon dioxide, as are the Earthlike polar caps observed by previous viewers. Giovanni Schiaperelli, who determined (erroneously) that Venus always keeps the same face toward the sun, discovered the famous (nonexistent) canals of Mars, launching nearly a hundred years of fiction about the advanced race of civilization there.

But despite the obvious similarities between Mars and Earth—the polar caps, the clouds, the twenty-four hour day, a similar chemical composition—the planets are vastly different. Mars is only one-tenth the mass of Earth, and this together with the absence of life probably accounts for the difference in its atmospheric composition, which in its proportions looks much like Venus': about 95 percent carbon dioxide, with very little oxygen and nitrogen. It is vastly different from both Venus and Earth, however, in total atmospheric amounts: there is only about 0.6 percent of Earth's volume of gases around Mars. There are probably two reasons for this, both dependent on the planet's small mass. Being so much smaller, Mars has never heated up internally like Venus and Earth, and so has not degassed as much; many of its gases are probably still locked inside it. And, again because it is so much smaller, it has less gravity and so has lost more of its atmosphere to space; on Earth only the lightest atoms such as helium are lost. The reason for the small mass of the planet, and for its density which is significantly less than those of Mercury, Earth, and Venus (3.97 vs. 5.2, 5.1, and 5.5 respectively) will have to be explained not by geologic or living processes during its history, but by the overall theory of the planets' formation which we are trying to develop.

CHAPTER TWELVE

•

THE MAJOR PLANETS

HALF A BILLION miles from the sun, five times further away than Earth, Jupiter is chilled by sunlight less than 4 percent as bright as Earth's. Water freezes there; anything liquid on Earth is frozen there on the outer atmospheric edge of a planet without a surface. We could recognize the other terrestrial planets as similar worlds, even if vastly different; but here on Jupiter we enter a new realm of existence.

Like the sun, Jupiter is 99 percent hydrogen and helium; like the sun, Jupiter is huge: with a diameter more than ten times Earth's, it is 1,400 times as big. It is bigger than everything else in the solar system put together, except the sun. Many stars in our galaxy are *binaries:* two stars orbiting each other, forming a binary stellar system. Our solar system is very nearly such a system; Jupiter is nearly big enough to be a star, but the gravity it exerts on itself is not sufficient to generate the temperatures and pressures in its core to begin the nuclear fusion process, and so it remains a cold planet. But actually not quite cold, either. Recent measurements show that it is radiating away more heat than it receives from the sun, so somewhere inside it is a source of energy. This is probably just gravitational energy released during its formation and trapped inside, slowly leaking out.

When Galileo first looked at Jupiter through his telescope, the first thing he saw was what looked to be a miniature model of a solar system, with the central planet surrounded by four orbiting moons. Better observations since then (primarily via spacecraft in the past two decades) have revealed twelve more moons, the orbits of the inner five in perfect alignment with Jupiter's equator, the others tilted at up to 33°, some of the outer

Photo 12.1. Collage of Jupiter and its four largest moons, photographed by *Voyager 1*. Assembled in proper relative positions, but not to scale. From upper left to lower right: Io, Europa, Ganymede, and Callisto.   JPL/NASA

ones with retrograde spin. The inner four moons are planet-sized, two of them closely bracketing Mercury's diameter, so the picture of a miniature solar system has some real validity.

Aside from the moons, the main features observed—and these can be seen easily through any decent child's telescope (better make that, through any child's decent telescope)—are the series of parallel bands covering the planet, and the Giant Red Spot (photo 12.1). These are due to a combination of Jupiter's gaseous composition and its fast rate of rotation: a Jovian day lasts less than ten hours, depending on where you are. The equatorial regions actually rotate faster than the higher latitudes, a fact made possible by the gaseous rather than solid composition of the planet. The white stripes, called *zones*, are atmospheric highs, similar to high pressure areas here on Earth, where warm gases are expanding and rising high into the atmosphere; there they finally cool off and spill over into the reddish stripes called *belts*, low pressure areas where they fall back toward the surface. Because of the rapid rotation of the planet, the equivalent of worldwide jet streams pushing the atmosphere at speeds of hundreds of miles an hour circle the planet and keep the zones and belts lined up parallel to the equator.

The Great Red Spot, so permanent that it was thought to be a surface feature, is now recognized as merely a hurricane; as a circus side-show barker might announce, it's the biggest hurricane in longest continued nonstop existence in the history of any world. It was first noticed by the English astronomer Robert Hooke in 1664 and by an Italian, Giovanni Cassini, in 1665. This was Galileo's century, when people first started looking through telescopes, so the red spot was probably there before that, and it shows no sign of going away, although it does move around the planet a bit. It's three times as big as Earth, and its winds reach velocities of 300 miles per hour.

In the 1950s radio astronomers were surprised, when they tuned their instruments onto Jupiter, to find emissions of radio energy. The first newspaper stories misinterpreted the scientists and reported radio broadcasts from Jupiter had been discovered—life on Jupiter! And it's just barely possible they may not have been completely wrong. The radio noise that was detected is intermittent and is due to lightning within the Jovian clouds, which are mixtures of hydrogen, methane, ammonia, and water. Just a few years before those first radio detections, two chemists at the University of Chicago, Harold Urey and Stanley Miller, carried out an experiment designed to investigate the beginning of life. They mixed into a flask gaseous solutions of water, methane, ammonia, and hydrogen, passed an electric spark through it, and found that amino acids—the molecules which are the stuff of which life is compounded—had been produced. The Jovian atmosphere is one big Urey-Miller experiment, with the same chemicals swirling around and with lightning sparks flashing through them. It's more than possible that amino acids have been produced there, and barely possible that these might have linked themselves together into some kind of primitive life; the temperature at the top of the clouds is − 113 degrees C, much too cold for life, and the central temperature is about 30,000 degrees C, much too hot for life, but somewhere in between the top of the atmosphere and the center of the planet we might find, when finally we get there, tiny living cells floating around in those thick clouds.

We haven't yet landed on Jupiter, but in the 1970s a series of *Pioneer* and *Voyager* spacecraft flew by to within a few hundred thousand miles and discovered that a magnetic field ten times as strong as Earth's surrounds the planet. Earth's magnetic field almost certainly arises from electric currents within our liquid iron core, but we don't think Jupiter has an iron core.

Actually we're not very certain about much of anything inside that opaque atmosphere. To determine something about the planet's interior we have to resort to mathematical models based on what we can see about its exterior. We know, for example, its size, and from its gravitational interactions in the solar system we can determine its mass and therefore its density. Matching this with spectroscopic observations of the chemical composition of the clouds, we can make guesses about its total composition, and that's how we arrive at the estimate that it's 99 percent hydrogen and helium. More detailed calculations indicate that its composition can be grouped among three components, analogous to the Aristotelian idea of Earth being composed of earth, air, fire and water; Jupiter is composed of gas, ice, and stone.

The gases are hydrogen and helium, the ices are water, methane, and ammonia, and the stones are metallic silicates similar to the stony minerals of Earth. Table 12.1 gives an estimate of the composition of the major planets in terms of weight percents.

Table 12.1

Planet	Gases	Ices	Stones
Jupiter	82	5	13
Saturn	67	12	21
Uranus	15	60	25
Neptune	10	70	20

But the terms "gas" and "stone" are used only as names, not as firm descriptions, because of the unearthly conditions on Jupiter. In the atmosphere and down to a depth of perhaps 15,000 miles or so, the hydrogen is indeed a gas. But at this depth the pressure becomes so great that the hydrogen is thought to exist in a form not seen on Earth, "metallic" hydrogen: not a metal as we think of the term, but simply a compound in

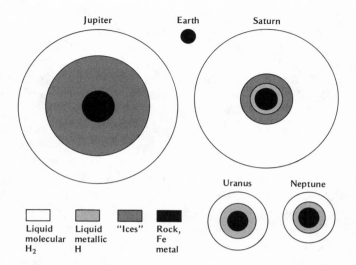

Figure 12.1.

JOHN WOOD

which the electrons flow freely from one atom to another as electrons do in such terrestrial metals as iron; in other respects the metallic hydrogen is thought to retain many of the properties of a gas, or at least a fluid. The outward flow of heat that we noted above keeps the inner zone of metallic hydrogen in convective motion, the planet's rotation keeps it spinning around, and such movement of a conducting metal (which is what it is) would generate the magnetic field. The central portions of Jupiter are so hot that the "stones" are also liquid rather than solid. A schematic diagram of the interiors of the major planets then looks like figure 12.1

Twice as far away as Jupiter, nearly a billion miles from the sun, is Saturn. When Galileo first looked through his telescope at it, he saw a strange, lumpy sort of planet. In 1655 Christian Huygens saw, in his improved telescope, what appeared to be "jug-handles" or ears sticking out of the planet, and within another decade these had been resolved into a set of concentric rings (photo 12.2). For three hundred years it was thought that Saturn was unique in having rings, and that there were three or four of them, but lately we have found similar (albeit smaller) rings around all the Jovian planets, and in 1979 when *Voyager 2* swung past Saturn after visiting Jupiter it found an uncountable series of rings, thousands of them,

Photo 12.2. Collage of Saturn and six moons, as photographed by *Voyager 1* and arranged not to scale. Shown are Dione (forefront), Tethys and Mimas (lower right), Enceladus and Rhea (left), and Titan (upper right). JPL/NASA

one after the other in chaotic profusion, providing the theoreticians with years of work.

The rings are not continuous sheets, but consist of innumerable small particles, dust-sized to pebble-sized pieces of water ice; they couldn't possibly be continuous sheets because, like everything else revolving around a central gravitation source, they obey Kepler's third law which tells them that the particles in orbit furthest from the source must rotate more slowly. So the inner particles move faster than the outer and so the ring, if it were continuous, would be continually pulled apart. Using this argument, Laplace suggested that instead of a single ring, the system consisted of a series of narrow, concentric, rigid rings each rotating at its own Keplerian rate. But Maxwell later showed that even this modification is not enough, for any small perturbation would shift a ring so that one side would necessarily come closer to Saturn; this would of course increase the gravitational force on it, which would bring it still closer, and the effect must feed on itself until the ring collapsed. So even a series of rigid rings would be unstable through time.

The striped pattern of Saturn is similar to but different from that of Jupiter, being less regular, less vivid, and of different colors, yellowish toward the equator and greenish toward the poles. This makes less likely the possibility that the coloration is caused by organic or biological molecules; crystallization of small inorganic molecules can also reflect various colors, and a slightly different chemistry and different temperatures could easily give the differences observed.

Uranus and Neptune are so far away, 1.7 and 2.5 billion miles respectively, that little is known about them. From Earth they are just two small, greenish-blue disks. But *Voyager 2* is still out there; it passed by Saturn in 1981 and on the last week of January 1986, it came within 51,000 miles of Uranus. In August 1989 it will encounter Neptune, we hope.

Before *Voyager 2* we knew virtually nothing about Uranus beyond its mass and density, its rotation angle, some of its moons, and the fact that it has rings. The planet (and Neptune also) is about fifteen times as massive as Earth, with the unexpected density of 1.6. This is much too large for them to be composed mostly of hydrogen and helium like Jupiter and Saturn, for although Jupiter's density is nearly the same (1.3) this is because Jupiter's immense mass condenses the light gases considerably; Uranus and Neptune aren't massive enough to do this. And yet the density is much too small for Earth-like materials; our silicate rocks have densities of about 3, and iron about 8. So Uranus and Neptune aren't gases and aren't stones; they must be ices. They are probably composed largely of the molecules that would freeze out at such large distances from the sun; water, methane and ammonia. These particular compounds are chosen because the elements carbon, nitrogen, and oxygen are the most abundant in the sun (except for hydrogen and helium), and these particular compounds are the most stable expected under the temperature and pressure conditions way out there where Neptune and Uranus fly. It is probably atmospheric methane which gives both planets their characteristic color when viewed through the telescope.

Photo 12.3. *Voyager 2* montage of the view a spacecraft would see skimming over Miranda toward Uranus.   JPL/NASA

Photo 12.4. Neptune as it might be seen from the surface of Triton, looking across a lake of liquid nitrogen. (Artist's impression.) JPL/NASA

At the present time we can't say much more about Uranus, and the perspicacious reader will have noted that most of what *has* been said is speculation. But this will metamorphose into hard facts as the current *Voyager* data come in and are analyzed. Particularly interesting is the detection of a magnetic field around the planet, analysis of which should tell us something about what is going on beneath the obscuring blue haze. (photo 12.3).

The rings of Uranus aren't actually visible from Earth. They were discovered in 1977 when a team of Cornell astronomers aboard a NASA plane high over the Indian Ocean prepared to measure the planet's diameter accurately by observing its time of transit across a star. As they were calibrating their equipment, they were surprised to see that the star blinked out a half hour before Uranus crossed it, and during the next thirty minutes it reappeared and disappeared five times, then duplicated this behavior immediately following the transit of the planet. The reason, they realized, was a system of five rings around Uranus, too tenuous to be seen from Earth but thick enough to block out the starlight as they passed between the star and Earth. By the time *Voyager* reached the planet it had detected ten such rings, most of them blacker than coal dust and presumably composed of either methane blackened by radiation or a carbonaceous tarlike material, but including two which reflect light differently enough to suggest a different (unknown) composition.

Uranus' spin is the most unusual in the solar system. Instead of spinning in the direction of its rotation, its axis is tilted about 95°, so that its axis of rotation lies nearly in the plane of rotation. During the summer seasons in the northern hemisphere the north pole is pointing nearly directly at the sun, and so daylight lasts for the whole summer as the sun merely revolves in a small circle overhead. And during winter of course, the opposite effect is seen. Since each of the seasons lasts over twenty (Earth) years, the effect would make for interesting vacations. Like Venus' retrograde spin, this motion is almost certainly due to an accident relatively late in its life, namely a violent collision between Uranus and a now-vanished planet-sized body. This postulated collision could also explain the Uranian rings, which might have been splashed out in the collision.

The five Uranian moons known from Earth observations have grown to fifteen with the *Voyager* sightings; all, together with the rings, orbit in

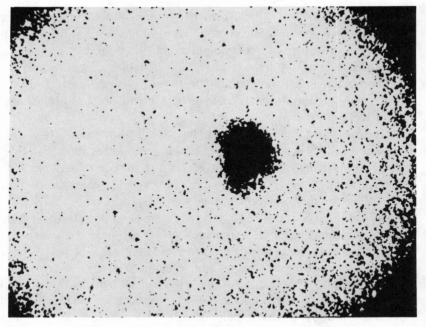

Photo 12.5. The best picture we have of Pluto. The elongation is Charon, incompletely resolved. U.S. NAVAL OBSERVATORY

the equatorial plane. The two Neptunian moons, by contrast, orbit well out of that plane, indicating they may have been trapped extrinsically. The largest moon, Triton, circles the planet backwardly (moving in a direction opposite to that of Neptune's spin), and is a source of much speculation. It seems to have a tenuous atmosphere composed of nitrogen and methane, with a surface composed of methane ice and with the distinct possibility of liquid lakes or even shallow oceans of molecular nitrogen (photo 12.4).

Pluto is so far away, nearly four billion miles, that we know practically nothing about it. We do know that it is much smaller than the major planets, but we don't know why. It may be that it is in essence a major planet that has lost all its light gases, or it may be a moon escaped from Neptune. In 1978 James Christy at the Naval Observatory noticed a strange, lumpy appearance of Pluto in the best photographs indicating a moon

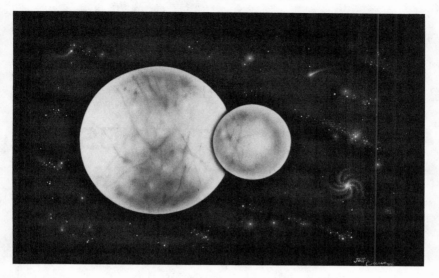

Photo 12.6. Artist's rendition of the meaning of photo 12.5.   JPL/NASA

incompletely resolved. (photos 12.5 and 12.6). It is possible that when Pluto was yanked out of its orbit around Neptune (if that ever happened) it might have shattered, creating its moon (called Charon) out of its own rib, as it were.

With Pluto we reach the edge of the solar system, so far as we know today. We now have to go back and fill in a few of the empty spaces, and then we can discuss how it all came to be.

CHAPTER THIRTEEN

•

MOONS, ASTEROIDS, AND METEORITES

A S THE PLANETS revolve around the sun, the moons revolve around the planets; and as the planets are of two major size classes, so are the moons. Seven of the solar system's forty-eight moons are about the size of the planet Mercury, while the remainder are much smaller. With one exception, these planet-moons orbit around the giant planets Jupiter, Saturn, and Neptune. The exception is our own moon, called by no other name, like a cat might be called simply Cat rather than Velvet or Bucephalus.

The moon, then, our moon, is a little over two thousand miles in diameter, orbiting us at a distance of a quarter of a million miles. It is barren of water, atmosphere, life, and any sort of geological activity— characteristics *not* shared with all other moons. The *maria*, so named by early observers who thought these dark patches were seas, are composed of dark rock called basalt and consist of lava flows; but they are all ancient, mostly produced when large foreign bodies impacted on the moon, tore through its crust, and allowed lava still molten from the moon's time of accretion to flow through onto the surface. A period of such intense projectile bombardments cratered the lunar surface, but ended nearly four billion years ago. A few impacts undoubtedly continued to occur, but none big enough or recent enough to be seen by man.

(As with every statement ever made about the moon, there are exceptions to this: medieval and early-modern accounts tell of transient spots, clouds, or flashes observed on the lunar surface. As one example, take the case of the Lost Crater: in Mare Serenitatis there had been observed, prior to 1860, a particular crater named after the naturalist Linnaeus. In 1860 the German astronomer J. F. Julius Schmidt reported that it had disappeared, and all that was to be seen in its former place was a hazy grey spot. Presumably the crater was obliterated between observations by a meteorite impact great enough to turn over the lunar soil but too small to leave its own visible crater.)

Before we went to the moon in the 1960s there were three competing theories for its formation. First is the fission theory, originally proposed by George Darwin (Charles' grandson), stating that the moon was ripped from the Earth as woman was born of man. The Pacific Ocean would be the hole left behind, the missing rib; it's quite a reasonable size for this. The second theory states that the moon was formed elsewhere in the solar system, was perturbed out of its original orbit, and then captured by the Earth. The third suggestion is that the moon accreted in orbit around Earth from rubble left over after Earth itself accreted.

Now that we have actually gone to the moon and brought back rocks for study, we can say for certain—not very much. In the first draft of this chapter, written in early 1984, I said the following:

> It can be said for certain only that the moon originated in one of these three ways, but which one we do not know. The chemistry of the moon rocks is similar to that of earth rocks, but still decidedly distinct; in particular, the lunar crust is not a good fit to the Pacific crustal rocks. This was at first a blow to the fission hypothesis, but it was then realized that if the moon fissioned before all the terrestrial iron-type elements were sucked out of the crust when the core formed, the match between Earth and moon is not too bad. Again, this similarity would seem to be a blow against the "elsewhere" scenario, but until we know the actual chemistries of the major planetary moons we can't argue against it. What we learned from the lunar rocks has provided stringent arguments which has led to the modification of each of the theories, but hasn't been able to verify or to eliminate any of them.

In fact the situation was a bit worse than this. In an article titled "Properties and Composition of Lunar Materials: Earth Analogies," published in 1970 in that most prestigious and correspondingly austere professional journal *Science,* Edward Schreiber and Orson Anderson of the Lamont-Doherty Geological Observatory in New York reported on lunar seismograph data obtained from the *Apollo 11* and *12* missions: "The sound velocity data for the lunar rocks were compared to numerous terrestrial rock types and were found to deviate widely from them. (However) a group of terrestrial materials were found which have velocities comparable to those of the lunar rocks." These are the data:

Samples	Sound Velocity
Sapsego cheese (Swiss)	2.12
Lunar rock 10017	1.84
Gjetost cheese (Norway)	1.83
Provolone cheese (Italy)	1.75
Romano cheese (Italy)	1.75
Cheddar cheese (Vermont)	1.72
Emmenthal cheese (Swiss)	1.65
Muenster cheese (Wisconsin)	1.57
Lunar rock 10046	1.25

Data for terrestrial rocks ranged from 4.7 to 10.8. The authors concluded that "perhaps old hypotheses are best, after all, and should not be lightly discarded."

Then in October 1984 a conference on the Origin of the Moon was held, in which most of the active workers were brought together to bang the ideas about a bit, and the result was mainly the discarding of all three hypotheses. (The cheese theory was not discussed.) Darwin's fission hypothesis demands a rapidly spinning Earth, since it has to be centrifugal force which flings the moon away, and no one can yet explain why the Earth was spinning so rapidly, why it has since slowed down, and how the moon's orbit has come to be 23° out of alignment with Earth's equatorial plane. The second theory, involving capture of the moon by the Earth, is physically unlikely and, despite many attempts in the past several decades to model it, remains so: another body might pass by the Earth or might crash into it, but the probability that its kinetic parameters would be such as to fit it precisely into stable orbit around the Earth simply boggles the imagination. It remains possible, but terribly unlikely. The third theory, accretion of Earth and moon from the same material, has become increasingly unlikely as we have learned to recognize the basic differences in chemistry between terrestrial and lunar rocks.

A ten-year-old but previously unaccepted theory was resurrected at the conference. Originally proposed in 1975 by William Hartmann and Donald Davis of the Planetary Science Institute in Tucson, it envisages a small planet grazing the Earth and knocking off a cloud of material which then aggregates into a moon. The impact would have vaporized roughly equal amounts of the Earth and the impactor; some of the material would then have fallen back to Earth, some would have flown off into space, and some would eventually have ended in a stable Earth orbit.

This model can explain some of the more obvious puzzles of the Earth–moon system, but perhaps only because since it is at the present time so vague, with parameters so numerous and values so unknown, it can't yet be pinned down. But it's the best we have right now. My own feeling is that the moon simply doesn't exist: perhaps the flat-earthers are right and the moon was invented by NASA to get money out of the government, just as Communism was invented by the CIA.

So we now have a total of five theories; at the present time I think it would be foolish to bet money on any of them. A central point of most of the theories, however, is that moon formation took place *after* solar system and Earth formation (if it took place at all), and so with a sigh of relief it is possible to claim that this seemingly insoluble problem has nothing to do with the formation of the Earth, and we'll leave it at that.

The two moons of Mars—Phobos and Deimos, from the Latin words for fear and terror, to go along with Mars, the god of war—probably aren't moons at all, in the sense of their origin. They were predicted long before they were observed, when Kepler said that since Earth has one moon and Jupiter has four, Mars must have two (double one for two, double two for four) which shows you the trouble with numerology, because although two Martian moons were indeed discovered in 1877, Jupiter now has sixteen.

Phobos and Deimos are tiny and oblong, with dimensions of only about 10 miles on a side, as we saw clearly when *Viking* passed by and took their pictures. From the surface of Mars they appear more like stars—small points of unresolvable light—than like our moon. Judging from their small size and nonspherical shape, they are probably asteroids captured from the nearby asteroid belt (as we'll see in a few pages, the asteroids are easily perturbed into Mars-crossing and Earth-crossing orbits).

The four great moons of Jupiter are called the Galilean satellites after their discoverer. A contemporary rival of Galileo, Simon Marius, claimed to have discovered them first; we reject his claim but accept his names: Io, Europa, Ganymede, and Callisto, mythological lovers of the great god. Toward the end of the nineteenth century a tiny fifth moon was found practically on the surface, less than 200,000 km up, named Amalthea for the goat that mothered the baby Jupiter, and then in this century eight more small outer ones were sighted from Earth. These are again named after Jupiter's women, the ones with names ending in "e" orbiting in a retrograde direction: Leda, Himalia, Lysithea, Elara, Ananke, Carme, Pasiphae, and Sinope.

Ganymede, the largest moon in the solar system, is larger than Mercury, and Callisto is only slightly smaller, but their densities are only about half that of the terrestrial planets, so they must be distinctly different sorts of objects. Io and Europa, on the other hand, do have terrestrial densities.

There wasn't much more than that known about the moons of Jupiter until NASA's spacecraft visited them, and we still know practically nothing about the four outermost moons except that they rotate retrogradely. Like the Martian moons, they may be captured asteroids. The next four closer in have highly inclined but properly directed orbits, and are made mostly of Earthlike rocks. Inside the next group of four, the Galilean moons, are a group of three tiny ones, Amalthea and two discovered by spacecraft, Adastrea and Metis.

The Galilean moons are worlds in their own right. Callisto, the outer one, is geologically dead, with a surface much more densely cratered than our moon. Its crust is covered with rock and ice, and its mantle is probably water ice. Ganymede, the largest and next in line, has had a geologically active past; it shows smooth plains and patterned grooves, with a relief varying by hundreds of meters, but there is no evidence of any volcanic activity within the past few billion years. Its surface is covered mostly with ice mixed with rock, but even with all this frozen water around, there is no discernible water vapor in the air—the temperatures are too cold, never getting above − 100° C.

Europa is the kind of thing kids make when they're in a vicious snowball fight: a ball of rock completely coated with snow and ice. Through the telescope it looks completely smooth and white; when *Voyager* got close enough to see it clearly, long curving cracks in the surface were seen, but no one knows what they mean. Perhaps the surface is a single ice sheet floating on a liquid water ocean, with enough heat coming up from below to shift and crack the ice. We don't know.

Io is the only moon in the solar system known to be currently geologically active. The theoretical prediction and nearly simultaneous observational discovery of Ionian volcanism is one of the most happy stories in planetary science. In the March 2, 1979, issue of *Science* magazine, Stan Peale, Pete Cassen, and Ray Reynolds of California published a series of equations describing the gravitational effect of the total Jovian satellite/planetary system on Io, an effect which forces the moon into an eccentric (elongated) orbit. They predicted that this forced eccentricity would lead to tidal heating effects (as it approaches and recedes from Jupiter) sufficient to melt a major fraction of its mass, and that a likely consequence of such internal melting would be violent volcanism.

Moons, as mentioned above, do not have volcanos. Not because they are different in kind from volcanically active planets, but because they are smaller. Being smaller, they cool off more quickly, and so any moon that might have been originally hot enough to generate volcanos is no longer so. A similar example is seen on Mars, where Olympus Mons was once an active volcano bigger than anything on Earth; but Mars, being so much smaller than Earth, has long since cooled off and the volcano is no longer active. No moon in the solar system is large enough to have held its heat sufficiently to remain volcanic for the four and a half billion years since it was created. But the mechanism discovered by Reynolds and his coworkers is a continuing, present-day one, and so they predicted on March 2, 1979, that when the *Voyager* spacecraft reached the Jovian satellite system its cameras might return "images of Io (that) may reveal evidence for a planetary structure and history dramatically different from any previously observed."

And just one week later, on March 8, 1979, *Voyager* came to within a few million miles of Io and took a picture that showed a cloud on the

Photo 13.1. Volcanic explosion on Io, seen by *Voyager 1* in 1979.   JPL/NASA

surface—but Io has no atmosphere. Further pictures and thermal emission studies showed that what they were seeing was a volcanic eruption spouting out SO_2 in an eruption bigger than any on Earth, and one by one another six volcanoes were spotted in the act of erupting (photo 13.1).

Despite the volcanos and its earthly density, Io and Earth are grossly dissimilar. The volcanoes are much more violent than anything on Earth, and the average surface temperature is −145°C, cold enough to freeze the erupting SO_2 clouds into a white solid. Io is a world covered by sulfur over a mantle of silicates. The tremendous tidal forces keep the interior molten and roiling, resulting in continual massive volcanism.

Christian Huygens, who discovered the rings around Saturn in 1655, found its largest moon, Titan, at the same time, but until *Voyager* flew past it in 1980 we didn't know very much about it. Its hazy appearance in Earth telescopes, suggesting an atmosphere, meant that we couldn't even measure its size accurately. We know now that it is the second largest moon in the solar system, approximately the size of Mercury. Its density is 1.9, corresponding to an equal mixture of rock and water, but with a temperature of only − 190°C all the water is frozen into ice. Nevertheless, Titan does have an atmosphere, the only moon to have one; in fact, its atmosphere is nearly twice as dense as Earth's, and consists mostly of nitrogen, as does our own. The minor constituents are quite different, though: instead of the 20 percent oxygen which our own planet has, due to the action of photosynthetic organisms, Titan's atmosphere contains a host of organic compounds such as methane, ethane, acetylene, and hydrogen cyanide. It may be covered by oceans of liquid ethane with dissolved methane, but we can't be sure as its surface is hidden from view by dense clouds of aerosols. Life under such conditions would be vastly different from anything living on Earth, but conceivably might exist.

The other twenty moons are covered with water ice and at least three of them (Tethys, Hyperion, and Iapetus) are 70–80 percent pure ice— gigantic snowballs more than four billion years old, with small rock cores. Titan is a roughly equal mixture of ice and rock, and the others are probably similar (perhaps about 40 percent rock and the rest ice). All except Enceladus are heavily cratered by impacts; Mimas, in fact, has one particular crater nearly as big as the moon itself. The unmarked face of Enceladus indicates recent melting, but the energy source isn't clear. The one remaining moon, Phoebe, is three times further away from Saturn than the others, wasn't seen by *Voyager*, and so we know nothing about it except that it travels in a retrograde orbit and may well have been captured separately from the others.

With the moons of the outer three planets we reach the limits of our observational knowledge about the solar system. The five large moons of Uranus, all rotating in the planet's equatorial plane and roughly the size of Saturn's (excluding Titan), were mysterious and vague until *Voyager* reached them just as this book was going to press, and now they are no longer vague—but they remain as mysterious as ever. Miranda, the inner moon (see photo 12.3) is made of methane and water ices intermixed with rock, but that is like describing Marilyn Monroe as a primate biped with occasional body hair. Miranda, like Marilyn, is much more than that: it is, as one NASA scientist described it, "a bizarre hybrid of every kind of exotic terrain in the solar system," with sinuous valleys like Mars, craters like the moon, inexplicable grooves and canyons rising higher than the Earth's Grand Canyon. Ariel and Titania are similar if less dramatic, and one might suppose that internal geologic activity induced by tidal forces from inter-actions with Uranus might be the cause, except that Umbriel (situated between Titania and Ariel) shows no sign on its cratered surface of any such internal stresses. The other ten moons *Voyager* discovered are much smaller, and we still know virtually nothing about them.

The Uranian rings are different from Saturn's in composition. The brightness of the Saturnian system is due to their water ice composition which reflects the dim sunlight well. Since the Uranian rings were discov-ered not by their reflected sunlight but only when they blocked out star-light, they must be dark lumps of carbonaceous material. Taken together with the dim reflectivity of the moons, this leads to the suggestion that they are similar materials, perhaps largely methane ice darkened by radia-tion damage over billions of years.

One of Neptune's two moons, Triton, is roughly as large as Titan. We suppose that it is composed largely of ices and rock, like the Uranian moons, but it is so far away from the heat of the sun that molecular nitrogen may have condensed there into liquid "oceans," as illustrated in photo 12.4. It moves in a retrograde orbit while the other known moon, Nereid, has the most elliptical orbit in the solar system. Such an unusual satellite system implies something different and violent happened to Neptune— perhaps the tearing out of one of its moons to form Pluto. If *Voyager* gets close enough, sometime in 1989, we'll learn a bit more.

About Pluto's moon we know nothing more than what can be seen in photo 12.5, except that its existence was verified in 1985 when it occulted the planet, and the dimming of reflected light as it passed between us and Pluto was observed precisely as predicted. It will probably be a long time before we know anything more about Charon.

The intense cratering observed on nearly all the moons is easy to understand in terms of a rain of small projectiles, centimeters to many kilometers in size, early in solar system history. The lack of an atmosphere allowed the projectiles to impact without burning up in the air, and the lack of geological activity preserved the craters through all time. Today we still see evidence of the tail end of such a bombardment in the hundreds of meteorites which fall to Earth each year. Because most of these are small creatures which vaporize in the atmosphere, and because most of the surface of the Earth is uninhabited by humanity (so that most of those that reach the surface intact go unrecorded), less than a dozen are recorded each year.

The history of meteorite falls through the ages is an interesting one, with a kicker at the end. The earliest use of iron by humans was initiated when pure iron "rocks" fell out of the sky, long before anyone knew how to mine, smelt, and purify terrestrial iron. G. A. Wainwright found iron beads in predynastic Egyptian tombs, strung together with gold. (And no wonder they were held in such value: in the dry climate they wouldn't rust as they do in our countries, and the only supply was at the infrequent and sporadic whim of the gods. A jeweler's dream.) The meteoritic origin of the beads was proven by chemical analysis in the nineteenth century, the high nickel content (7.5 percent) being typical of meteorites and unknown in terrestrial iron ores. Similar iron has also been found in the royal tombs at Ur in Babylonia, dating back to at least 3000 B.C.

The earliest recorded fall was at Crete in 1478 B.C., others were written about by Plutarch (705 B.C.) and Livy (654 B.C.), but the first intervention of meteorites in contemporary human affairs came about in 468 B.C. when the Thracians faced the Athenians across the river Aegospotamos in the Thracian Chersonesus. Anaxagoras, on the eve of battle, predicted a great Thracian victory. But the Thracians, bearing as they did an inferiority complex vis-à-vis the Athenians, couldn't quite find it in their hearts to believe him; whereupon Anaxagoras predicted that the gods, who were on *their* side, would surely (pretty please) provide an omen to assure the Thracians of their favor; whereupon a great fiery red rock came tearing "out of the sun" and fell upon the river with terrible thunderclaps and assorted noises; whereupon the Thracians let out a cumulative whoop and fell upon the terrified Athenians, who fled in dismay and terror. Anyway, so it is said.

The kicker to the story is that, although meteorite falls have been recorded through all history (the Great Stone in the Kaaba at Mecca is thought to be a meteorite, as is the one which fell in 1492 and is still hanging (so I am told) in the town hall of Ensisheim, and an inventory of a temple of the Hittite Empire which lists the geographical origin of their gold and silver, describes their iron as "from the sky"), nevertheless it wasn't until well into the nineteenth century that meteorites were accepted by scientists as genuine phenomena. Disbelief in the naive tales of savages and benighted ancestors carried over into the stories of one's contemporaries. Thomas Jefferson, on being informed that two Yale professors had in fact concluded that an unusual rock specimen must be a meteorite, replied that he "found it easier to believe that two Yankee professors might lie than that stones might fall out of the sky."

He was right, it *is* easier to believe. Irreproducible and occasional phenomena are difficult to verify, although the common sight of shooting stars—which are grains of dust burning up as they fall through the atmosphere—might have led him to visualize the possibility of falling objects large enough to survive atmospheric passage. Throughout the eighteenth century the fall of meteorites was not accepted as actual fact, despite the work of the Czech scientist E. F. Chladni, working at the University of Berlin, who published the first book-length study of meteorites in 1794. His work was greeted with a spontaneous outburst of apathy: nobody believed him, no one cared, except for a few scattered individuals in England and Germany who analyzed the suspected meteorites and found them to be different from terrestrial rocks, and who wondered about the notorious inefficiency of nature and the concomitant possibility of leftover crud from the Creation floating around in interplanetary space, occasionally by chance bumping into Earth as it wandered around its oribt.

And so a few people here and there kept the heresy alive, kept it talked about, kept people wondering, until finally in the spring of 1803 the French Academy of Sciences in Paris brought the subject up for open discussion, and as if in reply a whole shower of literally thousands of meteoritic stones fell on the village of L'Aigle, confirming for all time the reality of meteoritic existence.

Meteorites are of two basic types, stones and irons, suggesting to early workers their origin in a disrupted planet. Diamonds are occasionally found in the largest meteorites, and since these are high pressure minerals this again suggested an origin deep within a planetary-sized body, but about twenty years ago Mike Lipschutz and Ed Anders of the University of Chicago were able to prove that the diamonds originated in the shock pressures formed when such large meteorites hit the Earth, and today it is generally accepted that the meteorites represent planetesimals which never made it into full-sized planets—although an occasional unusual meteorite may be a chunk of the moon or even Mars, blasted off and sent into Earth orbit by a cometary or asteroidal collision.

The stone meteorites, in turn, are of two basic types, chondrites and achondrites. The chondrites are so named for their inclusions, chondrules, which are generally spherically shaped high-temperature minerals of unknown but probably primitive origins. Some of the achondrites, simply meteorites without chondrules, are igneous rocks similar to our volcanic basalts, evidence that they originated in planetesimals large enough to have experienced melting processes. Most meteorites, the chondrites in particular, have been unaffected by any such processes since their formation and therefore carry information to us from the beginnings of the solar system, information which has been lost from Earth by volcanic activity and the incessant weathering processes of the hydrologic cycle, and from the moon by the early bombardment and lava flows. One particular class, in fact, the carbonaceous chondrites, contain fragile organic molecules and water, evidence that they have never been heated to any significant extent. More chemical and isotopic information about the origin of the solar system has come from studies on these objects, which have been termed "the poor man's space probes," than from any other material we have.

An obvious source for the meteorites is the asteroidal ring, consisting of rubble of all sizes orbiting between Mars and Jupiter. Computer calculations have shown that particular groups of asteroids may well be perturbed out of their orbit by close passage to Jupiter, and the resulting scatter may take them into Mars-crossing or Earth-crossing orbits. The Mars crossers might hit that planet, or a close miss might again perturb them and throw them toward Earth.

Another possible meteoritic source is cometary: some groups of asteroids may be cometary nuclei from which the gases which form the brilliant tails have been lost. Whatever their immediate source, it is clear that the meteorites bring to Earth information about the earliest years of the solar system, information uncomplicated by later planetary processes—but terribly complicated in and of itself. It will have to be interpreted, as is all observational evidence, against a background of theoretical frameworks that will enable us to link together various arguments, suggestions, and possibilities into a coherent whole. We have enough of the observations at our fingertips now to be able to take a closer look at these frameworks. What, as they say in Hollywood, is the story line?

SOURCES: Kerr 1984; Morrison and Samz 1980; Nininger 1952; Peale et al., 1979; Schreiber and Anderson 1970

CHAPTER FOURTEEN

•

ROGUE STAR

T HE FIRST STORY of Earth, sun, and star formation that did not involve the gods throwing a ball back and forth or riding a burning chariot across the heavens or goddesses squeezing their breasts to spurt a shower of milky droplets across the night sky was the biblical story, and we know now that it's quite wrong. The heavens and the Earth were not created in the beginning, but rather the heavens existed for at least five billion years before Earth creation; the sun could not possibly have been created after the Earth, but rather the Earth was formed in some way out of solar material or interactions; there is no water above the firmament, and the waters below, which seemed so overwhelming to our ancestors, constitute only a small part of the total Earth.

It took some three thousand years for another theory to be born, which isn't really surprising: as we have seen, the Aristotelian/biblical picture of the universe was a sterile one and a dominant one. Until we had some real idea of how the universe was constructed—until, in particular, we knew the difference between the stars and the planets, until we knew what the solar system really looked like and what its relation to the rest of the universe actually was—there was no possibility of coming up with an acceptable theory for its creation.

The first scientific theory of Earth formation came in 1776, and was due to new and precise measurements of the shape of the Earth. It was found that the Earth is not a perfect sphere but is oblate: its equatorial radius is greater than the polar radius by just over 13 miles. To place any emphasis on this might be thought to be a recidivism to the time of the ancient Greeks, with their belief in the perfection of the divine circle, but in fact it is important. Gravity, which holds the Earth together, is a nondirectional force, exactly equal in all directions. A body accumulating under its influence alone will naturally form a sphere, and maintain that shape. The Earth, of course, is subject to an additional force: since it is spinning, centrifugal force is pushing it outward, but only in the plane of the equator. A fluid mass, subject to these two forces, must accept the shape of an oblate spheroid, bulging at the equator, exactly as the Earth has done.

But the Earth is not a fluid mass. Except for the small amount of water on its surface, the Earth is hard rock—so far as was known in 1776. And hard rock could not flow downward to the equator to give the observed bulge. The Comte de Buffon in that year suggested that the Earth must have been initially in a molten state throughout, allowing it to assume an oblate spheroid shape at the time of its formation. He noted that the existence of volcanos indicates that the interior is still hot, and calculated that the Earth would have cooled from an initial molten state to its current cool condition, with a few hot spots inside still giving rise to volcanos, in about 75,000 years, which he proclaimed as the age of the Earth. In order to provide an Earth initially molten, he suggested that it must have been torn from the sun, and as a mechanism he suggested a cometary impact. The scenario was that a massive comet struck the sun, great glowing masses of molten material were exploded out, some of them fell back into the sun and some of them fell into orbit around the sun, cooled off, and became the family of planets.

It was a good first try, and some aspects of his theory still are around today. But it didn't work in any sort of detail because of the composition of comets and the motion of the solar system. Comets are fantastic things, enormous but tenuous: the great glowing mass that we see is a collection of gases and dust so fragile that if the Earth passed through it we wouldn't even notice it. The Earth did, in fact, pass through the tail of the comet Halley in 1910; for months before the predicted event the streets of cities from London to New York to Tokyo to St. Petersburg were filled with prophets announcing the end of the world and the coming of the Messiah, and with more sober-minded gentlemen selling gas masks and shovels for digging shelters—harbingers of the prophets and salesmen of nuclear destruction still to come. And the transit of the Earth through the comet's tail came and went, and nobody noticed.

The head of the comet is more massive, but it is still a fragile, small thing by solar comparison. It is most probably a "dirty snowball," as suggested in the 1950s by Fred Whipple of the Smithsonian Astrophysical Observatory, composed of dust grains embedded in an icy matrix, quite insufficient to raise planetary masses out of a solar collision. Comets, in fact, must have been colliding continually with Earth just as meteorites have, but in all recorded history there is only one such event that hit with sufficient force to be recognized. In 1908 a brilliant fireball was seen crossing the sky over Russia, and immediately afterward seismographs worldwide picked up an explosion in Siberia, near the Tunguska river. Several years later the Russian Academy of Sciences sent out investigative expeditions; they finally found the site of impact, a place where the frozen tundra for miles around had been melted and trees were broken and bent outward from the focus of the blast. But there was no crater, as there should have been if a giant meteorite had hit, and the consensus now is that it must have been a fragile comet which broke up in the atmosphere just above the ground, releasing enough energy to melt the frozen surface water and break trees, but not enough to excavate the solid ground. (There are differences of opinion, of course. I believe the *National Enquirer* has uncovered hidden evidence that the site marks the landing of an alien nuclear-powered spaceship.)

Although comets vary widely in mass, there is no evidence for any comet ever attaining the size necessary to excavate planetary-sized masses from a solar collision. Buffon's theory died and was replaced with the Kant/Laplace hypothesis of nebular condensation, the material left over from the central condensation forming a ring or series of rings around the sun, and finally condensing or accumulating somehow into planets.

In this theory the disk around the sun was thought to contract under gravitational attraction to the central sun; as it did so it would spin faster due to conservation of angular momentum. As the angular momentum increased, the centrifugal force at the edge of the disk would also increase, until finally it would be greater than the gravitational force holding the outer portion of the disk firmly to the central sun. At this point the outer-most ring of the disk would be in essence blown away, separating from the rest of the disk which would then continue to contract, spin faster, and repeat the process at intervals. In this way the initial disk would be transformed into a series of concentric rings, until finally the remainder of the collapsing disk fell into the sun. The individual rings that had been left behind were then to coalesce somehow into the planets.

As the rings formed into planets they would be very hot, either because (in the early thinking) of a postulated initial hot state that was never clearly defined but was instinctively linked to the basic observation that somehow, magically, the sun is hot and therefore always was and therefore the initial disk must have been, too; or because (in later thinking) gravitational energy was released as the rings collapsed; or, finally, in Lord Kelvin's model, because of heat released as a cataclysm of "meteoric bodies" fell onto it in the process of planet formation. It was, in fact, on the basis of this postulated "hot earth" as an initial state that Kelvin's calculations of the age of the Earth depended. In 1897 his method had not yet been proven wrong by Rutherford's radioactivity, and was a distinct embarrassment to the new theory of evolution, for Kelvin's 25 million years of Earth history simply weren't enough for evolutionary processes to take place. The biologists, filled with a new sense of power and achievement by Darwin's magnificent synthesis of observations into understanding, threw off for once their customary and understandable feelings of inferiority with respect to physicists and rose up to protest Kelvin's limits to the time they needed: "Natural Selection will never be stifled in the Procrustean bed of insufficient geological time!" was sung to the tune of "We Shall Overcome" by hordes of biologists marching in the streets of London on their way to the 1896 meeting of the British Association, metaphorically speaking.

Kelvin took up the challenge. In 1897 he repeated and refined his arguments, showing with mathematical precision the inflexibility of the limits of geological time. By 1899 his speech was published in the United States, where it was read by Thomas Chrowder Chamberlin, a former Chief Geologist of the State of Wisconsin, then head of the geology department at the new University of Chicago; and Chamberlin was shocked. As a geologist he was well aware of the difficulties of judging reality from appearances, and had always maintained a healthy faith in the principle of alternative hypotheses: keeping many theoretical possibilities open while collecting observational data, rather than forcing the data into one pigeonhole. He immediately sat down and wrote a refutation to Kelvin's arguments, pointing out that while the mathematics might be exact, the assumptions underpinning the arguments were not. How did one know, he asked, that the Earth *must* form hot? Even if Kelvin should prove to be right in his assumption of a rain of meteoric projectiles upon the forming Earth, what evidence was there that they came in so fast that the heat they brought could not be radiated away as fast as it was delivered? In fact, he argued, the available geologic evidence tended to point the other way, that the Earth accumulated in a cold state. If so, all of Kelvin's complex mathematical reasoning, for all its sophistication and differential equations, was so much mumbo-jumbo.

At the same time a young Ph.D. in astronomy, Forest Ray Moulton, was finishing his thesis research at Chicago. His work, which had begun as a rigorous investigation of the Kant/Laplace nebular theory, ended as a detailed attack on the basic assumptions of that theory. It so took Chamberlin with its force that he began looking for an alternative hypothesis. Moulton's argument was that instead of collapsing in a series of concentric rings, as the nebular theory assumed, the disc material would spiral in toward the sun: the solar system should not collect into a series of concentric orbits. Furthermore, he found that as the collapse proceeded the central star should spin faster and faster—the angular momentum of the system should be normally fixed in the central mass, contrary to what is observed in our present solar system. He concluded that the nebular theory simply didn't work, and a completely new idea was needed.

Actually, not quite completely new. Chamberlin, with Moulton's help, went back to Buffon's idea, rejuvenating it with the too-small comet replaced by a wandering star. Since stars come in a large spectrum of sizes (see figure 9.2) it wouldn't be difficult to find one large enough to drive out tremendous blobs of solar material without being so large as to destroy the sun by collision.

By this time, however, the motions of the planets were known well enough to demonstrate the implausibility of any solar collision, stellar or cometary, giving rise to the solar system. A collision and resulting splashout would eject material in all directions, and if planets should form from the exploded masses they would drop into orbits circling the sun isotropically rather than all being in one plane, moving and spinning all in the same direction. The resulting solar system would look somewhat like our idea of an atom (figure 14.1a) rather than like the present solar system (figure 14.1b).

Moulton and Chamberlin therefore suggested that, instead of a collision, a grazing encounter might have occurred. In such a situation the

Figure 14.1.

gravity of the passing star could pull masses of solar material out toward itself; then, as it continued on its way, the solar material would trail after it, eventually spilling around into orbit around the sun. As the rogue star disappeared into the distance the solar material would coalesce into cold lumps which would in turn collide and fuse together into the planets. By the motion given the ejected solar material by the passing star, these planets would naturally be in orbit in the same plane; the sun itself would have been spun by the passing star so that it would also be spinning in the same plane, but their mathematical analysis showed that most of the system's angular momentum would in fact be gained by the family of planets, as is observed. Subsequent versions of this theory were proposed by Sir James Jeans and Harold Jeffreys in the first two decades of this century, and the stream of ejected solar gases evolved into a cigar-shaped ribbon which, upon breaking apart, naturally collapsed into bundles that then fell into orbit around the sun. This might account for the observed relative masses of the planets if the pulled-out solar material was thickest at the cigar's center and thinner toward both ends; this would give the massive planets Jupiter and Saturn in the middle, surrounded by the smaller planets toward and away from the sun. The theory is still being actively pursued at the present time by M. M. Woolfson at York University, but it has its problems.

First, temperatures inside the sun reach millions of degrees; material pulled out of such a hot environment would immediately be torn apart by the thermal motion of its own atoms, would explode into space and dissipate rather than condensing into planets. Woolfson meets this challenge by imagining either the sun or the rogue star to be in an early stage of its evolution when the encounter takes place; one of them, at least, is a protostar rather than a true star. In this stage it is really only a thickening cloud of gas and dust, perhaps hot enough to glow but certainly not hot enough to have ignited its fusion furnace: the internal temperature is not millions of degrees, and could well be cool enough so that material torn from it might condense. In this scheme the planetary material might come from either star.

A second objection is the probability argument mentioned earlier in this book. A passing encounter of the sort visualized must be exactly balanced in terms of the masses of the two stars, their relative velocities, and the distance of closest approach. Vary any one of these quantities and either insufficient material will be torn off or the two stars will be pulled together by their mutual gravity and collide. It's a very delicately balanced configuration that must take place, and it's Woolfson's triumph to have been able to show mathematically that such a balance can in fact be achieved, but its improbability in a galaxy of randomly moving stars is staggering. Woolfson's reply to this argument is simply that unique events are unique simply because they *are* so improbable, and yet they do occur. We are each of us witness to that. Consider the improbability of our own unique selves, each of us. I, for example, aside from the usual complement of arms, fingers, and toes, have a bland but pleasing exterior, a body weight of 165 pounds, blue eyes, a mole on my back, an ear for music but no voice for it, a weak stomach, a tendency toward cynicism and sentimentality, an aversion to blood sports and war—I could go on and on, as could anyone, describing in infinite detail the myriad characteristics that make each of us unique. Before my birth, what would have been the probability that a being with precisely these characteristics would emerge from my mother's womb? For each of us, before we occur, the probability of our occurrence is infinitesimally small. And yet we do exist, to no one's surprise. What *would* be surprising would be our duplication, our *dopfelganger*. A singular, *unique* event can occur no matter how improbable, but if it should occur twice we would have to begin to think that there must be some unknown process that renders it less improbable. If our solar system is alone in the galaxy, then its origin in an improbable event is not unlikely. If other such systems exist, however, we must look for more probable origins for them. The problem at the present time, then, is to find out if indeed we are alone in the galaxy or not. And today, in the early months of 1986, we still don't know for sure. Until we do, the improbability argument is not a strong one.

Much stronger, as the twentieth century grew from infancy into its formative years, was the eighteenth century geological argument of uniformitarianism, augmented by the philosophically similar concept of biological evolution. In both of these concepts things change, but slowly; in the Bible, remember, things happened suddenly and with the wrath of God. Just as more worldly children rebelled against the capitalism of their parents and flocked to Communism in the 1920s and '30s, the scientific children of the emerging generation rebelled against concepts of catastrophism which may have been subconsciously associated with the Bible stories they had been taught and subsequently learned to reject, and strived to return to a world in which all changes were slow and reasonable, in response to the steady pressure of fundamental laws operating in natural settings. This was undoubtedly reinforced by an aversion to thinking of ourselves as the center of the universe, as our ancestors had thought, an aversion to thinking of ourselves as the only living creatures in the universe. And probably, tinkling around the edges of our collective subconscious like hobgoblins and things that go bump in the night, was the simple fear of being alone in a vast, empty universe.

Back we went, then, to the formation of stars and to the possibility that planetary formation was a natural consequence of the processes involved, to a universe populated by uncountable stars each with their own system of planets revolving unseen.

SOURCES: Brush 1980; Burchfield 1975; Woolfson, 1984

CHAPTER FIFTEEN

•

THE INEFFICIENT UNIVERSE

THE PRIMARY OBJECT in the universe is a Main Sequence star: any objective observer, entering our universe for the first time and evaluating it, would come to this conclusion. These are not only by far the most common objects in the universe, they are beautiful, awesome things, burning brilliantly and constantly for millions or billions of years with the white-hot flame of a universal logic in which the loss in mass resulting from the fusion of hydrogen into helium is converted relentlessly into energy according to $E = mc^2$. When finally the process gives out, when too much hydrogen has been used and the star's core is mostly helium, it collapses and heats up again until the higher temperature at which helium can fuse to carbon is reached. Once again there is an outpouring of nuclear energy, the star expands, and becomes a red giant. But now we begin to see the inefficiency of all natural processes. At these high temperatures some of the nuclei are broken apart, spilling their constituent nucleons into the roiling star. These are protons and neutrons, and while protons are identical to hydrogen nuclei and are immediately lost among them, the neutron is something else. It begins a new kind of reaction, sliding into a nucleus, forming a slightly heavier isotope and, interspersed with radioactive decay and combined with successive generations of fusion reactions, new generations of heavier elements are formed until finally the star runs out of fuel entirely and collapses in on itself in a cataclysmic death throe. At this final stage, if the star is massive enough, it literally explodes and becomes a supernova. In the collapse process the center becomes a maelstrom of whirling nuclear particles, the nuclei previously formed being smashed and fused and forming an equilibrium mixture of the most stable nuclei, nuclei as complicated as iron or nickel with its 28 neutrons and 28 protons; and then finally as the supernova flashes and erupts, throwing everything out into space, the explosion creates hordes of neutrons from the nuclear reactions, neutrons which flash through the mix and slip into nucleus after nucleus, forming heavier and heavier nuclei, well beyond the bound of nuclear stability. Finally, as the cloud disperses and fades away into space, these heavy nuclei decay radioactively to form the heaviest isotopes of the heaviest elements.

The cloud drifts through space, mixing with other clouds, until finally one day it collapses again to begin the process all over again—only this time along with the hydrogen that forms most of its mass it includes the heavier elements, the silicon, iron, oxygen, aluminum and calcium and magnesium out of which a planet can be made.

That's the concept. It depends on a basic characteristic of our universe: inefficiency. This is a characteristic we see everywhere; in all things human, of course, but also in such things as stellar nucleosynthesis and sex and diamonds and penguins.

Snares Island penguins lay two eggs, but attempt to hatch only the second. The first, smaller egg will sometimes burst open of its own accord and emit its occupant, but there just isn't enough food for two chicks per mother, and the unfortunate little first-born thing dies. Nature is a wastrel, grossly inefficient. When humans mate, the male produces sixty million sperm in the hope that just one of them will fertilize the egg. The universe is the original Great Society, not concerned with husbanding its resources but recklessly throwing them around in wasteful profusion. This temerarious use of inefficient processes is not confined to living systems; in the purest diamond there are hundreds of billions of impurities. There is hardly a process in the universe that would be acceptable in any well-managed human factory or business.

It therefore seems reasonable that when the universe decides to build a star, this process too might be inefficient, wasteful. Of all the incredibly large amounts of mass that must gather themselves together and collapse into the forming star, we cannot expect that 100 percent will be wholesomely integrated into the stellar mass; some must necessarily be inefficiently used, wasted. What happens to this material? Most of it eventually falls into the formed star, and most of the rest is probably spun away, lost again to the empty depths of interstellar space. But some small percentage might fall into orbit around the star. This leftover stellar material forms the basis of the nebular condensation class of planetary formation theories, as it whirls around the forming central star and is spun by its conservation of angular momentum into a disk whence its slow accretion into a series of planets becomes possible. Planet formation in this scenario becomes a normal part of the creation of stars, due to the basic sloppiness of the universe in which unneeded and unnecessary isotopes and elements are synthesized in stellar interiors, blown out into space, mixed with primordial hydrogen, and collapsed again into new stars sufficiently inefficiently that planet-sized masses are left behind, whirling on their own axes.

According to this concept, we owe our existence on this lovely world to the impurity of the universe, to its inefficiency; we are truly the dregs of creation, the ashes of the stellar furnace, a sloppy afterthought in the mind of whatever god may be. At least that is the theory. Let's take a closer look at it.

First, let's review the observational evidence that must be explained by any theory.

1. More than 99 percent of the mass of the solar system is in the sun, but

2. More than 97 percent of the angular momentum is in the planets.

3. The planets all move around the sun in the same direction and rather well within the same plane (except for Pluto), and

4. They spin all in the same direction, close to the plane of the sun's equator (the ecliptic), except Venus which has a slight retrograde spin and Uranus which is tilted at right angles. We don't know Pluto's spin.

5. Most of the moons also spin in the same direction as their planets, close to the plane of their planet's equators. Some of the outer moons of Jupiter have retrograde motion, as does one moon of Neptune.

6. The planets can be divided by mass and density into the terrestrial and major planets, the four terrestrial planets being the closest to the sun. Among the terrestrial planets there is no pattern of mass versus distance, but there is a gradual diminution of density moving outward from the sun. Pluto does not fit into this scheme.

7. The terrestrial planets are composed mostly of metal silicates and

iron, the major planets are composed mostly of hydrogen and helium. Uranus and Neptune have higher densities than Jupiter and Saturn, and seem to have more carbon, nitrogen, and oxygen, in the forms of methane, ammonia, and water.

8. The planets exhibit a fairly regular spacing of orbits.

9. The terrestrial planets have 0 to 2 moons each, while the major planets have up to 21.

10. All the solid bodies of the solar system which have been measured—earth, moon, and meteorites—are 4.6 billion years old, formed within about 0.1 billion years of each other, i.e., just about contemporaneously.

11. The chondritic meteorites, aside from their loss of volatile elements such as the rare gases, have elemental abundances virtually identical to those of the sun. Isotopic abundances, aside from well-known radioactivity or cosmic ray effects, are nearly but not quite exactly the same in all solar system bodies we have examined. ("Nearly but not quite" is one of those dangerously innocuous phrases that can hold hidden a host of devastating anomalies; remember it).

The original twentieth-century rejection of the nebular condensation theory was based on two points: James Clerk Maxwell had calculated that all the mass of the solar system, if spread out in a nebular disk, would not have sufficient gravity to pull itself together into separate and distinct planets, and Moulton and Chamberlin had calculated that a solar disk would not naturally form the set of concentric rings that Laplace had visualized, and that most of the angular momentum of the system would end up in the central mass (sun). The stellar encounter theory they replaced it with, however, is so intrinsically improbable that it could be accepted only if we are virtually unique in our galaxy. This idea, with its philosophical undertones that necessarily return us to the pre-Copernican concepts of an anthropomorphically centered universe, was appealing to some and distressing to others. Such subjective emotional criteria, of course, have no place in science—and yet they have their inscrutable place in scientists, they influence our behavior, they twist us and pull us along their paths.

And so the idea of solar systems that form through natural stellar formation processes, necessarily resulting in a universe populated throughout with planetary systems in wild profusion, would not die. The drive to find such a solution stayed alive, even during the terrible British and American bombing of Germany during the Second World War, and during the height of that cataclysm, in 1944, the German scientist Carl von Weizsacker resurrected Laplace's nebula. He made one important change: he suggested that the process of condensation from a cloud into a central star is not only inefficient, but *grossly* inefficient. He was the first one to free us from the observation that the sun constitutes 99 percent of the mass of the solar system. (The problem with observations is that unrecognized assumptions trail along in their wake and sneak through into our theories like submarines in the old movies, lurking under warships and then sneaking through dangling nets into harbors.) The observation that the solar system is to a very good approximation simply the sun itself had led everyone to postulate a process where that necessary conclusion was a condition from the outset of the process, but now von Weizsacker evoked a model in which the left-behind nebula was not a thin, tenuous membrane but rather a thick web with many times the mass of the current planets in it, so that its gravity would be sufficient to overcome Maxwell's objection.

Two physical properties act upon such a rotating web. Friction between the nebular particles would tend to homogenize their periods of rotation around the central sun, making them all rotate in a common period of rotation; but Kepler's third law, a consequence of the inverse-square centrally directed force of gravity, tends to make the inner particles move faster than the outer. The former property would make a ring that rotates as a unit, the latter creates tidal stresses that break it up. Von Weizsacker claimed that the final result of the clash of these two forces would be a bipartisan breaking up of the thick nebula, the inner portions falling into the sun and slowing its rotation while the outer portions would be broken up into a series of vortices. The triumph of the theory was that if one addressed the configuration of the vortices in a center-of-mass frame of reference (that is, if one drew a diagram of what the system would look like not to an outsider but to an observer situated at the center of mass), the geometry of the situation necessarily dictated that as each system of vortices accumulated into a planet it would sweep up all nebular material within a region such that each succeeding planet was roughly twice as far from the sun as its inner neighbor—and this closely approximates the Titius-Bode law.

At the same time, and not too far away, and also totally unbeknownst to scientists in other countries, the Chamberlin/Moulton objection was being dealt with. This objection, you will recall, was based on the supposedly unalterable and inescapable results of the conservation of angular momentum.

An interstellar cloud moving through the galaxy is naturally going to be tumbling. Think of any irregularly-shaped object thrown into a pool of water: the transition into a region of higher density must occur first at some particular point on the object which will be slowed down while the rest is still moving fast, and this must result in a spinning or tumbling effect. The same effect will be felt by the cloud as it encounters higher density spiral arms or feels the blast of a supernova; certainly as it begins to collapse it will be spinning. And no matter how slowly it is spinning, as it collapses its spin must increase due to conservation of angular momentum. The change in linear dimension is tremendous; it has collapsed from an initial radius of about 10^{15} miles (a million billion) to only a billion miles, a decrease of a factor of one million, and its angular velocity must increase proportionately in order to keep the angular momentum constant. But angular velocity is reflected in centrifugal force pushing material outward from the spinning protostar, and as the velocity increases, this tremendous centrifugal force becomes sufficiently great to overcome the gravity of the mass and destroy the protostar by flinging its material away into space. In other words, as the cloud condenses it spins faster and faster, to the point where it self-destructs.

And yet it doesn't. Stars exist, which means that protostars must. Somehow this angular momentum must be diminished in the process of star formation.

The first successful attempt at answering this point was by a young Swedish electrical engineer, who brought a new point of view to an old problem. He suggested that the magnetic field of the collapsing cloud, although only one-hundred-thousandth the strength of the Earth's, might be sufficient to dissipate the growing angular momentum. The concept can be understood by considering how we generate electric power from coal or water or nuclear reactors.

Figure 15.1 shows a simple loop of wire placed between two magnets, with a galvanometer connected across the two ends to measure the flow of current that will be generated.

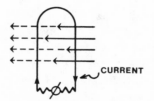

Figure 15.1.

The magnets induce a magnetic field between them, which can be thought of as a number of magnetic lines of force (shown as arrows) connecting them, passing through the wire loop. Faraday's Law of Electric Induction states that if the strength of the magnetic field inside the loop is changed, an electric current will flow through the loop. The field can of course be altered by changing the strength of the magnets, but it can also be changed simply by moving the magnets or the wire. If, for example, the wire is moved vertically out from between the magnets, the lines of force will no longer pass through the loop: the flux will have gone from its initial value to zero. In an electric generator or dynamo the movement of water may be used to turn or move the wire, or coal may be burned to provide steam power to move it, or the steam may be created by nuclear energy release in a reactor: whatever the initial source of energy, it is converted into mechanical motion of either the magnets or the wire, and this results in a change in the magnetic flux through the loop and therefore creates a flow of electric current through the wire.

In such an electrical generator, what is being accomplished is merely the conversion of the energy stored in the coal or water or uranium nuclei into electrical energy—the electrical energy is not being created out of whole cloth. More energy must be dissipated in the process than is recovered as electricity; if it were otherwise a small amount of coal could light our cities and we wouldn't have to rape the countryside and pollute the atmosphere in order to get enough. But it is not otherwise: this unpleasant but inescapable fact is the consequence of the law of conservation of mass-energy, which says that mass-energy can neither be created nor destroyed, and of the principle of entropy which is here to assure us that no real process in this all-too-real universe can be accomplished with 100 percent efficiency.

All very well—we have been brought up to suffer principles gladly—but what does this mean in actual physical terms? Consider the coil of figure 15.1: a certain amount of energy X is necessary to move it in and out of the magnetic field at a certain rate Y which will induce a certain flow of current Z. Now imagine a universe in which the coil is floating freely between the magnets, with no force acting on it and no friction to hinder its motion. Then obviously only an infinitesimal force would be needed to start it moving, and only infinitesimal increases would be necessary to get it moving at the rate Y; that is, in order to provide the electrical energy indicated by Z, much less energy than X would be needed—only an infinitesimal amount of energy would be needed to provide any finite electrical energy.

Lovely. Unfortunately, such an imaginary universe is not the one in which we live. In our all-too-real universe there is a force acting on the wire as a consequence of the magnetic field in which it sits, and in order to move it, to generate electrical energy, this force must be overcome, which means that work must be expended, coal or nuclear or some other form of energy must be used, and consequently there is no such thing as a free lunch.

Actually, this situation is not *too* unfortunate—because if such an imaginary universe existed, we would not. It is this precise no-free-lunch situation, this juxtaposition of the laws of our universe, which brakes the spin of the collapsing interstellar clouds and allows stars and planets to exist. The only difference is that the emphasis is reversed.

Any kind of motion of the wire loop in figure 15.1 will induce a current flow. Imagine that we do this by spinning the wire. The resulting change in magnetic field strength through the loop induces current flow in the wire; in a similar manner the flow of current induces its own magnetic field. The two coexisting magnetic fields try to align themselves, but since the wire is spinning its induced magnetic field is also spinning and cannot align itself with the stationary field of the fixed magnets: this creates a force between them that will act in the direction of slowing down the speed of rotation of the wire. In an electrical generator more coal is burned to provide more mechanical energy to keep the wire spinning at a constant rate and keep the electricity flowing; in a collapsing interstellar cloud there is no more coal to be burned, and so the spin must slow down.

This model of a spinning wire in a magnetic field is valid not only for a wire but for any conductor of electricity, and despite appearances the interstellar cloud is a good conductor. I say despite appearances, because we have spoken of the cloud as a low density atomic or molecular gas which would not seem to be an efficient conductor, but the constant impingement on it of starlight and cosmic rays means that some of the atoms are ionized: there will be in the cloud free positive ions and free negative electrons, and electricity is defined as the flow of electrons. In this ionized gas the flow of electrons is facilitated precisely because of the low density: they can move through space with little likelihood of collision. And so a contracting cloud of such ionized gas, beginning to spin rapidly because of its conservation of angular momentum, is a very good approximation to a rotating wire and the weak galactic magnetic field replaces the two magnets of figure 15.1, with the same result. An electric current is induced in the cloud, carried by the moving ions and electrons, and this in turn generates a magnetic field which locks against the original galactic magnetic field and stiffens the collapsing gas against further spin. The result is that angular momentum is transferred out of the cloud into the surrounding intergalactic medium, and the speed of rotation of the cloud is slowed below the point at which the forming protostar at its center would self-destruct because of overwhelming centrifugal pressure. Voila!

Regrettably, the magnetic concept did not appear with sudden and dramatic flair, but rather slipped in through a half-open door. As far as I can tell, the first suggestion of the existence and importance of this process came in a largely unnoticed paper, one that couldn't have been better hidden from the world if Nixon himself had stonewalled it. The paper, "Remarks on the Rotation of a Magnetized Sphere with Application to Solar Rotation," was published by a young Swedish electrical engineer, Hannes Alfven, in *Avk. Mat. Astron. Fys.*, and whatever one says about the internationality of science, the Swedish language is as good as the largest haystack ever thrown together if one wants a place to hide golden needles in. The timing wasn't bad either: the year of publication was 1942, when half the scientists in the "civilized" world were trying to kill the other half, working on the atom bomb or radar or trying to find submarines hidden in the Bay of Biscay or making proximity fuses or studying the logistics of armies traveling on their stomachs, when Germany was driving deep into the guts of the Soviet Union and Japan was sweeping across Asia and half-dead Americans were marching across Bataan with their hands raised.

These two concepts—the massive solar nebula and magnetic braking of angular momentum—are probably the biggest advances in our understanding of the origin of the solar system since Newton's concept of gravity. They rank with Friedrich Wöhler's laboratory synthesis of urea in the nineteenth century as one of the all-time unappreciated giant leaps of mankind, dwarfing Neil Armstrong's dramatic first step on the moon.

In 1828 at the University of Göttingen Wöhler changed forever our concept of life by heating a flask of ammonium cyanate, carefully monitoring it, and showing without doubt that it had rearranged its molecular structure and become the organic molecule urea. For the first time a "biological" compound had been created in the laboratory. Since he had first created the ammonium cyanate from the basic elements carbon, nitrogen, hydrogen, and oxygen, it meant that synthesis of biologic compounds from elements was possible, which in turn removed the concept of magic from the "miracle" of life; from that moment on it became possible to think of the life process as a chemical one, life became amenable to scientific research, and not only the synthesis of such organic compounds as penicillin and aspirin became possible, but the manner in which we think of life became altered and all things became possible.

In precisely the same manner von Weizsacker's concept of a massive nebula, together with Alfven's magnetic braking and transfer of angular momentum, changed the entire direction of our thinking about our origins. They allowed us to think of planetary formation as a natural process in the life of a star instead of a unique event, an ad hoc catastrophe, a miracle.

Both ideas were temporarily lost in the turbulence of the war, but in the following decades they stimulated a host of new investigations into the possibilities of nebular theories. The regularity of von Weizsacker's vortical patterns was eventually shown to be unlikely, so the essence of his model appears unworkable, but it brought the scientific community back to the nebular class of hypotheses, and when in 1954 Alfven incorporated his idea into a book published in English by the Oxford University Press, the idea began to spread. (In 1970 he won a Nobel Prize for related work on magnetohydrodynamics and plasma physics.) Fred Hoyle, sharer in the concepts of the steady state universe and the stellar nucleosynthesis of the elements (for which, as we have seen, he did *not* win a Nobel Prize) picked up on this magnetic coupling idea in 1960, working out mathematically how the sun might have transferred its original angular momentum to the primordial solar nebula and thus spared itself centrifugal destruction. And with this the way was opened to return to the concept of Kant and Laplace, which had been demolished by Moulton and Chamberlin on the basis of their quite correct but (as it now turned out) incomplete angular momentum argument. Once the Moulton/Chamberlin demolition had been undone, in fact, various people came up with a variety of methods to disentangle the Gordian knot of angular momentum which had strangled the nebular theory: it now became possible to transfer angular momentum from the sun to the disk by turbulent mixing processes or by simple friction due to differential rotation rates. It was as if once Alfven had shown us that the Medusa was only a shadow, suddenly everyone had a singing sword and no one could remember what all the fuss had been about. So it goes.

To recapitulate: Laplace had developed a detailed model of the Kant scheme, ignorant of the implications of angular momentum. In this model he proposed that a spinning protosun would generate a nebular disc around its equator and would shed consecutively concentric rings of material to later coalesce into planets. Moulton and Chamberlin at the beginning of the present century showed that the angular momentum of such a system would result in the contracting central sun spinning faster and faster, so that the outflung nebular disc would be dragged behind and would take on a spiral shape rather than that of a set of concentric rings. Clearly such a situation would not result in planets moving in concentric orbits. Their argument had such force, was so unanswerable, that at one stroke it clouded men's minds toward the nebular concept and redirected their thoughts toward the necessity of catastrophic and therefore intrinsically unlikely origins. But once their unanswerable argument had been answered, once the possibility of magnetic coupling to dissipate the angular momentum of the central mass and transfer it out to the disc material had been demonstrated, we were free once again to return to a planetary system growing in the normal course of events, naturally and easily, as a sideshow to the main event of the birth of a star.

Over and over again we see this in history: a previously unassailable conclusion proven wrong by a point no one had dreamt of before. The lesson is clear: nothing is certain, all things are possible, dreams are the stuff of which our future is made. As the Prologue of *The Invisible Ray*, starring Boris Karloff and Bela Lugosi, put it in 1936:

> "Every scientific fact accepted today once burned as a fantastic fire in the mind of someone called mad."
> "Who are we on this youngest and smallest of planets to say that the INVISIBLE RAY [here one might substitute "nebular condensation theory" or "polio vaccine" or "fusion energy" or any one of a multitude of projects past and present] is impossible to science?"
> "That which you are now to see is a theory whispered in the cloisters of science. Tomorrow these theories may startle the universe as fact *(sic)*."

To put it a bit more prosaically, as Ted Ringwood of the Australian National University has suggested, we should be hesitant about throwing away hypotheses on the basis of single lines of argument, no matter how strong that argument seems to be at the time, for who knows what new insights the future may bring? All of which is not to say that the stellar condensation problem has been totally solved, any more than Wöhler's synthesis of urea solved the ultimate question of the origin of life. It is simply to say that the paths of attack now being taken *feel* right; there is confidence that ultimately an answer will be found at the core of the onion whose layers are being progressively stripped away. The overwhelming though not unanimous consensus of workers in the field is that miraculous, once-in-a-galactic-lifetime catastrophes are not necessary to create this world we live on.

The problems remaining are qualitatively different from the situation at the beginning of the century, when the angular momentum problem was an absolute one: unsolvable in principle. These problems are—or seem to be, one hopes—the sort of problems that science is meant to deal with, the sort that are in principle amenable to further investigation, which we will deal with in the remaining chapters.

SOURCES: Brown and Mussett 1981; Simpson 1976

CHAPTER SIXTEEN

•

STELLAR CONDENSATION

THE BEGINNING of the process can properly be put at the moment that a heterogeneous mixture of gas and dust, the interstellar cloud, begins to condense. The particular cloud that was to form our sun and planets almost certainly was originally part of an immense cloud, thousands of times more massive than the sun. The collapse of this giant cloud was not a quiet, orderly affair, resulting in a superstar at its center; rather it was a tumultuous, turbulent process in which the cloud fragmented again and again, forming perhaps thousands of smaller clouds, each of which would finally condense into denser aggregations and become opaque to infrared radiation.

This marks the end of the fragmentation process, because what has been happening is that as the clouds condense they release gravitational energy, just as water releases energy as it flows over a dam. This gravitational energy is undirected, unfocused by any governing mechanism, and in this universe such wastage of energy means automatic conversion into heat. The inescapable less-than-100 percent efficiency of all mechanical processes, and the consequent loss of energy to heat, is the anathema of our industrial society: our car engines cannot make 100 percent use of their fuel, and in the process they heat up and need complicated mechanisms to cool them down, and nuclear reactors get unbearably hot because of this wastage, requiring incredible quantities of water to flush them out and keep them from melting (and when this coolant water is returned to the ambient environment it is hot enough to affect it adversely, resulting in the spontaneous generation of pickets and legislation).

In the collapsing star there is no problem at first, as the heat generated is efficiently radiated away into the infinite reservoir of the universe; the wavelength of such heat radiation is in the infrared region, and such wavelengths escape the dilute gas easily. But as the cloud continues to condense, its density increases finally to the point where it begins to absorb the radiation, and from this point on the heat generated by its collapse can no longer escape the cloud. It heats up, its internal pressure rises as the heat is shared among the gas particles in the form of rising velocities, and the increased gravitational pressure is balanced. The rate of compression slows and stops, the temperature rises to the point where the whole mass is glowing, and the process temporarily stops.

At this point the object is roughly the size of a solar system rather than a star, and its surface temperature is about 4000° Kelvin; we are talking here about an object that will eventually become a star of solar mass. This surface temperature is not much less than that of our present sun, and so it is glowing brightly—much more brightly than the sun, in fact, for with a temperature not much less it is very much bigger—and is in fact what Lord Kelvin envisaged as a star. It is not, though; we call it a *protostar* because this process of gravitational release of energy could only continue, for a star the mass of the sun and burning with its present intensity, for about 20 million years (as Kelvin calculated). When the internal temperature reaches a few million degrees, the H→ He fire is ignited and we begin the Main Sequence stage: it is now a true star.

The condensation of each of these individual clouds into stars, while more orderly than the collapse of the initial giant cloud, would be no less inefficient: as each tumbling cloud contracted, its speed of rotation would have to increase in order to maintain a constant angular momentum, and so instead of a sphere at its center the cloud would form a flattened spheroid with centrifugal pressure pushing it outward along the equator. The final effect would vary depending on the conditions of collapse. Sometimes the result would be the formation of a dumbell-shaped central mass which would split finally into two lumps revolving around a common center, and here we would be forming a binary star system. Other times the central mass configuration would be even more complex, resulting in a ternary system or one even more fragmented. The most fragmented system of all would be one with a number of equally massive bodies revolving around each other, but this doesn't seem to be a dynamically stable evolutionary system: the most likely result is either two fairly massive central bodies or even three; otherwise we tend to end up with just one central body and a larger number of small bodies orbiting around it. And that, of course, is just what we want.

It's always pleasant when a probable result of our theories gives us just what we want; it's even more pleasant when we can see that sort of thing actually happening. Luckily, we can see a good deal of this hypothetical process happening right now. We might have expected that star formation would have been restricted to an early and specific period of the universe, but that's not the case. The sun, as we have seen, is four and a half billion years old; its total life expectancy is about ten billion years, and very few stars live longer than that. Probably every star we see is younger, which means that since the universe is roughly 15 to 20 billion years old they did not form early on, soon after the Big Bang. Rather, star formation seems to be a continuing process throughout universal history; for example, the best information we have about a star cluster like the Pleiades indicates that the entire cluster is only about 60 million years old. In fact, the simple fact of a star cluster's existence is evidence that the stars that form it are relatively young: in our scenario, the cluster is formed by the collapse of a massive cloud, fragmenting as it collapses and each fragment condensing finally into an individual star. The resulting cluster is gravitationally bound, as the original cloud was—weakly. The stars on its periphery in particular are quite weakly bound. They begin to feel the gravitational tug of the rest of the galaxy nearly as strongly as that of the cluster, and in time will begin to wander off from the cluster just as the outer molecules of a drop of water are less tightly bound by intermolecular forces than are the inner molecules; thus the drop slowly evaporates, and so too does the star cluster.

From age estimates of stars within visible clusters we infer that the maximum lifetime of a cluster is just a few billion years, and so the fact that we see so many clusters today (tonight) is ample evidence that star formation and cluster formation is an ongoing process. Incidentally, this conclusion fits well with our own sun: we are not part of a stellar cluster, but since we are over four billion years old we would have expected our original cluster to have evaporated by now. We can, however, see new star clusters in the process of forming—or at least we can see, by using both our eyes and instruments which can detect electromagnetic radiation at wave lengths which are invisible to us, objects both precise and nebulous which we interpret most reasonably as stellar clusters midway through their birth process.

The night sky appeared to the ancients to consist only of black void together with the moon and the stars. Later they learned to differentiate between the stars and another class of objects, the planets; as they continued to observe they began to see other things: stars falling out of the sky, comets presaging great events such as Caesar's murder, and finally in 1572 they saw the sudden appearance of a new star, Tycho Brahe's supernova. (The supernova of 1054 whose remnant still exists today as the Crab Nebula was recorded by the Chinese, but Christian Europe at that time still would have burnt at the stake anyone who suggested that the heavens were not eternal and perfect, and so no European appears to have seen it. So much for the objectivity of scientific observations.) Today we understand all these astronomical sightings, and our sophisticated instruments have brought a host of wholly new heavenly objects to light (so to speak).

The Orion Nebula (photo 16.1) is a cloud of gas and dust about 15 light years in diameter, just over a thousand light years away from us. It is lit by a group of young stars within it. It is dense enough (about 1000 atoms per cubic centimeter) to scatter the starlight and thus become visible; normal "empty" interstellar space has a density of about one atom per cc. The brightest central regions of the nebula are even denser; the outer regions, extending into darkness, do not drop off abruptly into normal interstellar space but diminish gradually. If we measure the radiations emitted by particular gaseous molecules such as CO and H_2, we can follow the existence of the cloud out to several times the area of the optically visible regions. By using infrared detectors we can determine that the central bright regions are intrinsically more bright than they appear to be: much of the visible light from these regions is obscured by a high density of dust.

This dust is sometimes itself visible in a negative sort of way, as shown by the Horsehead Nebula (photo 16.2), where the concentration of dust grains is so high (in the shape of the horse's head) that it shows up as a black region in front of the nebula it obscures. Such dust absorbs the starlight and reirradiates it at infrared wave lengths instead of in the visible spectrum; the clouds therefore appear dark to our eyes, but as "bright" sources to infrared detectors. In the central bright regions of the Orion

Photo 16.1. The Orion Nebula.   LICK OBSERVATORY PHOTOGRAPH

Photo 16.2. The Horsehead Nebula.   LICK OBSERVATORY PHOTOGRAPH

Nebula we see just such a group of infrared sources, indicating the presence of dust there—insufficient to blot out the light behind it because it has contracted to small, discrete, blobs instead of remaining as a "fog." These small infrared sources are probably stars in formation, not yet hot enough to turn on their hydrogen-helium fusion furnaces. Although our resolution isn't good enough to see if there are planets forming around them, they are an excellent illustration of the first stages of the proposed condensation sequence of star formation. (A Catch-22 in the process is that stars naturally form in localizations of high density dust, and so the dust obscures the details of formation from us: where the dust density is low and the seeing is good, no stars are forming. *C'est l'astronomique.*)

One might reasonably express a polite doubt at this point: is it, in fact, at all reasonable that stars should not only have formed at some incomprehensible time in the past (we must accept that they did, for we exist) but have continued to form by the processes we envisage *in an expanding universe?* If we think of the universe as the still-expanding product of an immense explosion, as we do, how can we expect portions of that expanding mass to come together in clouds and condense to form stars in such universal profusion?

The first point to note in answering that question is that stars do *not* form in profusion, but only very infrequently throughout space and time. Our sun is in a typical region within a typical galaxy, and if you look in any random direction from here your gaze will pass through hundreds of light years—thousands of billions of miles—before it is even likely to encounter a star. The very closest stars are a few light years away; a sphere drawn around the sun with a radius of 10 light years would contain less than ten stars. Put another way, the average density of stars in a galaxy like our own, the Milky Way, is about 0.002 stars/cubic light year: in a typical cubic box one light year on a side, there would be on average only 0.002 stars—you would need to search roughly five hundred such immense boxes before you would be likely to find a single star. And in the universe as a whole, the galaxies are distributed even more sparsely.

The conclusion is obvious: the universe is mostly empty space. Stars, although immediately impressive because of their bright appearance against the dark background, are minor components—it is the dark empty background itself that is typical of the universe we live in. That stars do occasionally form can be understood by considering the basic physics of the gaseous clouds. The atomic density of interstellar space is about one atom (or molecule) per cubic centimeter. At this incredibly low density (many orders of magnitude more "empty" than the best vacuum ever produced on earth) the starlight that pervades the galaxy is unhindered, undiminished by absorption; it heats the interstellar particles to a temperature of actually thousands of degrees. (It would be wrong to think of the virtually empty space they inhabit as being this temperature, although on earth we normally think of the temperature of a room as identical with the temperature of the gas molecules within that room; the difference is that space is so empty, so devoid of these hot particles, that they fly through it without interaction.)

Within the tenuous interstellar clouds the particle density is roughly fifty times as great. This is sufficient to diminish significantly the incident starlight by absorption, and so the temperature inside such a cloud is much lower, perhaps 50 to 100 degrees absolute. So we have a cloud with much lower internal heat than its surrounding medium, and therefore its atoms have much lower thermal velocities: when they collide they push each other apart with much less pressure. In addition, the greater density means greater gravitational force pulling the cloud together. At its normal density this gravitational attraction is not enough to begin collapse, or even to hold the cloud together were it not for the hot pressure of the surrounding medium pressing in on it, but the cloud is gently attracted to the galactic plane by interaction with the galactic magnetic field. As many such clouds slide into the galactic magnetic valley they inevitably collide, merge, and form bigger, more massive, more dense clouds. By the time the cloud has grown to about a thousand solar masses, with a density on the order of a hundred or more atoms per cc, its gravity has increased and its internal temperature has decreased enough for it to become gravitationally instable; it is on the razor's edge, waiting for some external disturbance to induce collapse.

What sort of disturbance is this likely to be? Until recently it had been thought that random motion of the molecules within the cloud itself would eventually bring enough particles close enough to begin gravitational contraction, and once begun the attractive forces would increase as distance decreased so the effect would snowball and proceed to star formation. But it now seems clear that something more specific than random motion and eventualities is operating.

A star the size of the sun will live for about ten billion years. Will stars more massive, with more hydrogen to use as fuel, live proportionally longer? The answer, perhaps surprising at first glance, is precisely the opposite: because of their greater release of gravitational energy massive stars burn at a hotter temperature, and since the fusion reaction rate is sensitively affected by temperature they burn their fuel at a prodigious rate. They can race through the main sequence in a few million instead of a few billion years. Consequently, the massive stars in our galaxy are all youngsters compared to the sun, and we have recently come to realize that these young, massive, bright stars are not distributed randomly or homogeneously throughout the galaxy. They are seen to be concentrated along the leading edges of the spiral arms, which is precisely what makes the spiral arms as bright as they are.

An illustration of our galaxy is shown in photo 16.3. This is actually a photograph of the Andromeda galaxy, but its characteristics are similar to

Photo 16.3. The Andromeda Galaxy.   LICK OBSERVATORY PHOTOGRAPH

our own Milky Way. The central bulge, although in the picture it seems to be a single glowing mass, is composed of individual stars, each of them light years apart from each other. They are not resolvable because of the galaxy's great distance from us and because of the great number of stars in it, somewhere on the order of a hundred billion.

Spiral galaxies like Andromeda and the Milky Way are obviously revolving, but they do not revolve as solid bodies, with each star in the galaxy retaining a fixed position with respect to all the other stars. We see this clearly in galaxies outside our own, and within our own Milky Way we observe the effect by measuring the Doppler shift in the light of intra-galactic stars: some of them show a blue shift, indicating they are coming toward us, others show a red shift which indicates movement away from us. Studies of these apparent stellar velocities with respect to us allow an estimate to be made of their motion and ours through the galaxy. We find that the sun, located about one-third of the way out from the center, is moving at a velocity of nearly six hundred thousand miles an hour around the center of the galaxy. Comparing this with our distance from the galactic center (about 30,000 light years) gives a period of galactic rotation for us of roughly 250 million years. The periods of other stars are different, depending on their distance from the center.

This differential rotation is responsible for forming the spiral arms, which are composed of a high density of stars and interstellar matter revolving around the center, being twisted around it by the differential velocities as they revolve. But by the same token the spiral arms should be ephemeral features, twisted into oblivion by being stretched into cir-cularity nearly as fast as they are formed. And yet this is not so; if it were, spiral galaxies would be unusual objects in the night sky. On the contrary, although some galaxies appear globular and some irregular, the spirals are the typical galaxies that we see: obviously they are of some permanence. Our own galaxy, in particular, should have had its spiral arms disappear within a billion years or so, and yet we know that it is more than ten billion years old and the spirals are still going strong.

The clue to the solution of this problem is the observation, mentioned above, that the massive young stars are concentrated in the leading edge of these spiral arms. Such stars last only a few million years—less than a hundred million, certainly—before fading and disappearing from our view. That is, they will not last through one complete galactic rotation, and yet the spiral arms which they form do last for many rotations, for billions of years. The arms must not be permanently composed of such stars, then; which must mean that they are not permanently composed of any material at all. Instead they are spiral density waves, composed of regions of higher density that sweep around the center. Then as an interstellar cloud passes into such a wave it will be suddenly encountering a region of higher density, and this is likely to initiate gravitational collapse of the cloud and formation of stars within it, and lead naturally to the observation of young stars concentrated within such a region: a region of high density, a spiral arm of the galaxy. As the wave moves on, leaving the new stars behind, they quickly die and fade into the darkness of the interspiral regions.

A variation on this theme is the idea that condensation of the proto-solar cloud may have been initiated by a supernova explosion, supernovae being predominantly located along the same spiral arms. But whether it is a density wave or a supernova shock wave that began our solar system, we have recently come to realize that the passage of our sun through the architecture of the galaxy may have important consequences not only for solar system formation but for the continued history of life on earth.

Life on earth, as all of us know (as *nearly* all of us know) has developed from simple to more complex creatures through evolution. The history of life on this planet shows continual disappearance of species and appearance of new species: life forms evolve or die out, and their ecological niche is taken over by new creatures. The classic picture of evolutionary development is one of slow changes throughout geologic time, with many millions of years necessary for one species to slowly evolve into another. However, the geologic record includes comparatively sudden "events" in which an abrupt change in the fossil record is seen. The most famous is the boundary between the Cretacious and Tertiary periods, roughly 65 million years ago, which is defined by this sudden change in life as seen in the fossil record. The most dramatic change is the disappearance of the dinosaurs, which had been the dominant animal group on earth for some 200 million years. They disappeared overnight, geologically speaking, and no one knows why or how.

Over the years, many explanations have been offered: a change in weather or in the food supply, a genetic disease, an influx of cosmic rays, a virus like the influenza virus which killed millions of people seventy years ago, a change in sea level—the list goes on and on. Recently a serious advance has been made in our understanding of this extinction event by two new insights. First, it is now clear that the extinction of the dinosaurs was not an isolated occurrence. At or nearly at the same time many other species and genera became extinct, from microscopic plankton to swimming fish. Second, Luis and Walter Alvarez and their coworkers at Berkeley have found a worldwide layer of iridium precisely at the Cretaceous-Tertiary boundary.

The significance of the iridium layer is that this is an element severely depleted on the surface of the earth; most terrestrial iridium is in the core. But meteorites have the normal cosmic abundance, and so it becomes a marker for meteoritic events: measurements of iridium around a proposed impact site will show enhancements over the average terrestrial value if indeed a meteorite did strike at that time. The Alvarezes have found just such an iridium enrichment precisely at the Cretaceous-Tertiary boundary from a dozen sites around the world, and they interpret this as evidence that a monster meteorite (or asteroid, or comet) struck the earth at just this time. It's not clear exactly what effect this would have on our environment, but it is clear that the effect would be tremendous. One possibility is a gigantic dust cloud that would shield the entire earth for tens of years, cutting off photosynthesis and thereby destroying plant life as well as lowering temperatures cataclysmically. Such effects could well disrupt the food cycle and cause widespread extinctions.

While the details are still being argued about, it has been generally accepted that some extraterrestrial event must be responsible for the iridium layer, and the coincidence of its occurrence precisely at the Cretaceous-Tertiary boundary strongly implies that the same event is responsible for the mass extinctions that occurred at that time. Furthermore, it now appears that such extinctions are not terribly scarce in the geologic record. Although the dinosaurs are the most prominent victims, similar extinction events have occurred before and since, characterized in the geologic record by the sudden and simultaneous disappearance of large groups of fossils and followed soon afterwards by the emergence of evolutionarily new groups. The evidence is spotty, but it does look as if these events are not randomly distributed in time; they show a periodic reoccurrence (see figure 16.1).

The cycles occur roughly every 26 to 32 million years, and this coincides very well with the periodicity with which our solar system passes

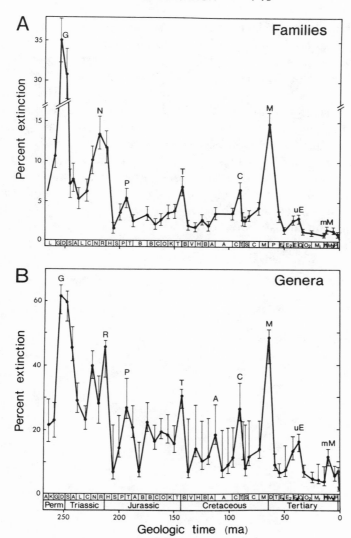

Figure 16.1. Periodicity of Mass Extinction   DAVID M. RAUP

through the galactic plane. That is, the sun in its galactic revolution continually rises above and sinks below the plane of the galaxy (where the mass of the galaxy is concentrated). In these peregrinations it passes through the central plane with roughly the same periodicity seen in the extinction record, and various suggestions have been made as to possibilities in which the galaxy might interact with the solar system: clouds of interstellar gas could perturb our cometary cloud, sending a barrage of comets against the earth, or disruption of the asteroidal orbits might throw them against us, or a deluge of supernovae might wrack us with high energy particles and blast waves. None of these are clearly proven, but one way or another the synchronization of extinctions with our passage through the galactic plane may be significant. Thus, as in our personal lives, the conditions of our cosmic birth may return time and again to influence and change the direction of our past and future development, in ways undreamt of by the astrologers.

SOURCES: Habing and Neugebauer 1984; Herbst and Assousa 1979; Kerr 1984; Scoville and Young 1984; Stothers 1985

CHAPTER SEVENTEEN
•
OBSERVING THE NEBULAR DISK

A S A STAR begins to form by gravitational collapse from the interstellar cloud of gas and dust, falling into the protostar stage on a
time scale of something under a million years, the centrifugal outward-
directed force of its spin along the equatorial plane directs its development
not into a sphere but rather into a central flattened spheroid connected to
a nebular disk. This is the crux of our hypothesis on the formation of
planetary solar systems, and indeed we see evidence for it in astronomical
studies of the infrared stars mentioned previously. These are stars which
are not observable in visible light because they are surrounded by a cloud
of dust, but as the dust is heated by the central fire it reirradiates in the
infrared region. Many such stars are embedded in young star clusters, and
are thought to represent protostars. In the mid-1960s one of these stars, FU
Orionus, suddenly began to get brighter: in four months it increased its
brightness by more than two orders of magnitude. This must be due either
to the collapse of a protostar into the hydrogen-burning stage with a
consequent flaring up of energy, or to the flattening of a circumsolar shell
of dust into a disk, removing most of the central star's obscuration. And
this is not an isolated astronomical occurrence. Many stars, particularly
young and variable ones, are associated with infrared radiation surrounding them, almost certainly representative of an absorbing disk of dust.

The most recent and spectacular observation began in the fall of 1982, when it was noticed that Epsilon Aurigae began to darken. This is a normal, unexciting third magnitude star—the fifth star (epsilon) in the constellation Auriga, just above Orion in the night sky—which one night began to fade away. By the end of the year it was half its original brightness. It then stabilized for a year and gradually began to brighten again, returning to its normal state in another few months.

Actually, it began long before that. Epsilon Aurigae's darkening did not just "happen" to be noticed in 1982—people were watching for it. It had happened four times before, occurring at regular intervals of 27.1 years ever since it had first been seen in 1904 by Hans Ludendorff in Germany. At that time the basic principle was clearly explained, and it seemed interesting but not exciting. The dimming is caused by an eclipse: the star is actually a binary pair, but we see only the bright binary. When the smaller, less luminous companion in its revolution around the larger star comes between it and us, the dark companion necessarily partially eclipses it, and the total light output to us naturally dims.

Such eclipsing binaries are not particularly rare; after all, most of the stars in our galaxy are binary pairs, and these will be seen as eclipsing in all cases where we happen to lie in the plane of their revolution. The problem with Epsilon Aurigae was the long time-span of the eclipse: to dim the light for eleven months, as it does, the companion star must be very big—and if it is so big, it should be bright enough to be seen.

Twenty years previous to the 1982 cycle, Margharita Hack, director of the Trieste Observatory, had proposed that the companion was a truly massive star, therefore a very hot, young star, and that it wasn't seen because it radiated mainly in the ultraviolet rather than the visible region of the electromagnetic spectrum. It wasn't possible to test this model immediately, because ultraviolet radiation is almost completely absorbed in the earth's atmosphere, but when the International Ultraviolet *Explorer* satellite flew in 1978 as a joint venture of NASA and the European Space Agency the experiment suddenly became possible, and just a few months after launch the characteristic UV radiation was observed: the companion is indeed an ultraviolet star.

But it is also an infrared star. Measurements made during the eclipse by the infrared observatories at Manoa in Hawaii and at Kitt Peak in Arizona showed significant *if* radiation from the companion, and a detailed comparison of the results at the two different wavelength regions have given a precise picture of a star being formed, surrounded by a shell of gas and a disk of dust. The main cause of the Epsilon Aurigae eclipse is a disk of dust particles which absorbs the light of the primary star as it crosses in front of us. In the center of the disk is the nearly invisible ultraviolet star, heating the disk so that it glows in the infrared. Enveloping the star and the disk is a shell of gas. The ultraviolet observations show that the companion star is itself varying in brightness, and this is due not to any sort of eclipse but to the nature of the star itself: its characteristics place it in the class of stars called Herbig variables, named after the California astronomer George Herbig, who explained them years ago as young stars in the process of forming from surrounding clouds of dust and gas. Because of the nature of the binary system in Epsilon Aurigae, the dust grains within the gas cloud around the young, still-forming star, have formed a nebular disk, just as we expected.

Developments in the techniques of infrared and ultraviolet astronomy in the 1980s have provided increasing instances of what appear to be stellar systems in the process of forming planets. In 1983 the Infrared Astronomy Satellite (IRAS) was launched by NASA as a joint project operated by the United States, England, and Holland. To calibrate the infrared telescope aboard, it was aimed at Vega, a young star known officially as Alpha Lyrae, the fifth brightest and, at 26 light years from earth, one of the closest and most studied stars in the sky. It has sightly more than twice the mass of the sun and is 50 times as luminous in the visible wavelength region, but the IRAS observations showed it to be much brighter in the infrared than it should have been. After a series of observations which eliminated the unexciting possibilities, such as an unknown galaxy which might be lying just behind it and adding its own infrared illumination to Vega's, the only thing left is a visibly invisible source (i.e., a source of infrared which doesn't emit visible light) and that means a cloud of dark particles around the visible star, absorbing light and emitting infrared. The size of the dust particles in the cloud can be guessed at from the particular wavelengths at which it radiates: they appear to be roughly gravel-sized. But larger sizes might well be present and unobserved; for example if all that mass were in the form of one single planet we wouldn't see it with the experiments that have been run. The total mass of the cloud is estimated from the total amount of radiation emitted, and comes out to be about 1 percent of our earth—much smaller than our solar system as a whole. But remember that we are seeing only one particular size class of particles. If the cloud has what we regard as a normal mass distribution—similar to, for example, what we see in the asteroid belt—the total mass of the cloud comes out to be about 300 times that of the earth, just about the mass of our solar system. The only thing unusual about this, from the point of view of what we might expect from our theory, is that the solid particles form what appears to be a spherical cloud rather than a flattened disk. The explanation is probably that it is really a disk, but we happen to be looking at it face on. And that explanation fits in well with another observation, that no rotation has ever been seen in Vega—which can only be explained if Vega's axis of rotation is pointed directly at us, for there's no way to form a star without setting it spinning.

A few months later IRAS found the same situation around another young star, Fomalhaut, and then in 1984 an even clearer picture emerged from three large terrestrial infrared telescopes at Kitt Peak and Mauna Kea, and a new technique, *speckle interferometry,* which helps remove atmospheric blurring from the data. In a normal stellar photograph, taken with a long exposure time in order to gather enough light from a dim object, the picture is made up of many individual "speckles" of light, like a painting by Seurat. Each individual speckle is perfect, and in the absence of a turbulent atmosphere the total composite photograph would be just as perfect. But our atmosphere is unfortunately turbulent, and its effect is to smear one speckle over another, resulting in a less clear picture.

In the technique known as speckle interferometry, thousands of pictures are taken, each of them less than a hundredth of a second in exposure. Each picture produces its own "speckle," and then each individual speckle is broken down into its component frequencies by a mathematical technique known as *Fourier analysis* and then all of them are put together again with the aid of a computer, giving a picture of the star as it would look in a long-exposure photograph without atmospheric turbulence. And in these new data we seem to be seeing precisely what we expect. Two young, obscure stars, 100,000-year-old HL Tau in the constellation of Taurus, 500 light years away, and one-billion-year-old R Mon in Monoceros, 2,000 light years away, are surrounded by dust which is clearly in the form of disks. Our solar system is about 40 AU in radius; the HL Tau disk is about 100 AU, the R Mons disk somewhat smaller. The total mass around HL Tau is comparable with that of our own solar system, while R Mon's is about the same as our earth. And in just the past few months IRAS has found disks around two more stars, and a collaborative group from Massachusetts, Wyoming, and Hawaii have found ground-based evidence for one more, this last one a star so young (Lynds 1551/IRS 5) that it's still buried in the dark molecular cloud from which it formed and is visible only in the infrared. The IRAS stars are roughly the same size and mass of the sun but are ten times more luminous, which is precisely what we expect from a solar-type star caught in the act of birth. (The release of gravitational energy in the collapsing protostar leads to a final prestellar burst of luminosity just before its final collapse into the H-burning main sequence stage.) And finally, these objects are nowhere near unique; although IRAS operated for less than a year before dying out, it found nearly twenty other objects that closely resemble these stars aborning. Two improved infrared satellites are scheduled for launch in the 1990s, and more sophisticated experiments will be able to look for more precise details.

Normal visual astronomy, meanwhile, is trying to keep pace. It would be lovely to actually *see* all this happening, and we keep trying so hard and being so clever that new results are announced continually; this section must be out of date by the time it gets printed. The most recent development as I write this is that Bradford Smith of the University of Arizona and Richard Terrile of the Jet Propulsion Lab in California have announced visual sighting of a disk around one of the IRAS infrared sources.

The problem with visible light, in the case of a star surrounded by a disk, is that the central star is so much brighter that the disk is lost in the background. To get around this inconvenience, they adapted a technique from the method used to measure the light emitted by the sun's outer envelope, its corona. The same problem applies here: the sun's corona is too dim compared to the central disk to be seen in normal times, so measurements are taken during an eclipse, when the central disk is blocked out by the moon. It gets to be too frustrating waiting for eclipses, so coronal workers designed a coronograph, a device which mimics an eclipse by blocking out instrumentally the central bright light. Smith and Terrile adapted this instrument at the Cerro Las Campanas Observatory in central Chile, and pointed it at Beta Pictoris, one of the close-by IRAS sources, a young star (100 million years old) twice as bright as the sun. With the

Photo 17.1. A circumstellar disk of material around Beta Pictoris, probably composed of ices, carbonacous material, and silicates, photographed by Bradford A. Smith and Richard J. Terrile. JPL/NASA

central disk blacked out and the most sensitive light detectors looking just outside it, they finally saw (not too clearly but unmistakably) what our solar system must have looked like nearly four and a half billion years ago, shown in photo 17.1.

These very solid infrared and visible data have certainly established the reality of gas and dust disks co-forming with new stars, and the rapidly increasing number of these observations indicate that these are not born of exceptional circumstances. Such observations go a long, long way toward establishing the general concept of a normal evolutionary sequence as the origin of the solar system, rather than relying on a unique catastrophic event such as a stellar encounter. The theoretical prediction that the combination of conservation of angular momentum and magnetic interaction will result in a collapsing cloud forming both a central mass and a surrounding disk is abundantly confirmed, and we can go on to the next step with confidence that we're on the right track.

The next step, however, is a tough one.

SOURCE: Boss 1985; Hack 1984

CHAPTER EIGHTEEN
•
ACCRETING PLANETESIMALS AND DISSIPATING PROTOPLANETS

THE ABUNDANT observational evidence showing the formation of disks around young stars is a strong message that we are on the right track in developing our solar system as a natural consequence of the formation of the sun. But the next step in the process, the segregation and accumulation of the disk into planets, is a difficult one to model—made easier to imagine but more difficult to pin down by the total absence of any hard evidence. We see interstellar molecular clouds and supernovae and stars and stellar disks—all the material that goes to form the planets—and we see our own planets four and a half billion years after their creation, but nowhere have we ever seen planets in the process of formation, or disks in the process of aggregating into anything at all.

The various theories that have been proposed to model the process can be grouped into two distinct classes, one in which the buildup proceeds from accumulations of dust into larger and larger bodies, and one in which original gravitational instabilities collapse to form giant protoplanets which then strip down to the planets we see today. The latter class of theoretical models has been constructed largely by Al Cameron of Harvard

University who, following previous work by Lust in Germany and Lynden-Bell and Pringle in England, has envisaged a gas and dust disk that grows with time as material settles gravitationally into the equatorial plane of the protosun, and once there is spun outward because of the centrifugal force. In about 50,000 years the disk will have accumulated two solar masses of material and grown to a size of five or six hundred AUs. At this point the outer edge is spun off into space and the inner regions collapse onto the sun, so that the disk contracts to an outer limit at about the present orbit of Neptune and a total mass not greatly above that now seen in the planets.

The question now is whether such a disk will clump into discrete bodies, and Cameron suggests that while no simple gravitational instabilities will occur in which direct condensation into planets would be possible, a different type of instability will manifest itself. He calls these "axisymmetric instabilities," which means that they will form not discrete bodies but rings around the nebular center. The formation of such rings depends upon the mass of nebular material in each region; taking "regions" as defined by the locations of the present-day planets, he has found that there should indeed have been sufficient mass in each region to form separate ring instabilities. Different mechanisms of ring generation have also been presented by A. J. R. Prentice of Australia.

Once these rings have been generated, the matter in them must collide and form symmetric gravitational instabilities which will grow into a set of independent globs of gas (remember most of the material of the disk will be hydrogen and helium), and as these globs in turn collide they grow into giant gaseous protoplanets, ranging in size from about the present size of Jupiter to perhaps thirty times as large, and all at distances well over 1 AU from the protosun. The inner rings formed first, according to Cameron, last according to Prentice. Let's follow Cameron for a while.

The inner rings formed so quickly in this model that much of the solar cloud of material had not yet fallen into the plane of the disk. As protoearth formed, further material was gravitationally attracted to it and the protoplanet grew; as it grew its gravitational attraction to the protosun grew apace, and it spiraled inward. As it grew and spiraled in, the heavier dust grains that it was sweeping up fell through its hydrogen-helium atmosphere toward the center, forming a rocky core. This continued until protoearth came close enough to the protosun so that tidal forces induced by the protosun's gravity swept the gases from its surface, leaving behind a small rocky protoplanetary core in a stable orbit. A similar history would have occurred for all the inner, terrestrial planets.

The protoplanets that began life further out in the nebula never got close enough to the protosun to feel the destructive tidal waves sweeping away their outer layers. They continued to accrete material until finally their own gravity reached the point where they collapsed in upon themselves. As they did, conservation of angular momentum made them spin faster, spinning their material out to form a disk around each of them, similar to the protosolar disk out of which they formed. These protoplanetary disks then shed material to form satellite systems just as the protosolar disk formed the planets.

This protoplanet approach has been formulated and championed by astrophysicists like Cameron and Prentice. Cosmochemists, on the other hand, have gravitated toward a totally different picture of planetary formation, in which dust particles successively accrete and grow by stages until the planets are complete. I don't mean to imply that physicists and chemists don't talk to each other; quite the contrary is true, and both physical and chemical data are constantly mixed into the two groups of models, changing the models and directing their development as new experiments are performed and old ones are refined. But the impetus behind the protoplanet models are astrophysical calculations, classical mechanics in a protoplanetary setting, while the driving forces behind the planetesimal theory have been more on the chemical and petrological side.

In this model attention is focused on the solid dust grains rather than on the gas. These fall through the gas cloud toward the equatorial disk, pulled there by gravitational forces. (Centrigugal force creates the disk, which at first is only a bulge—a compression of gas and dust. This bulge has a slightly greater gravity than the surrounding spherical cloud, and pulls the dust particles toward it; their motion, perpendicular to the centrifugal vector, is not inhibited by it.) As the dust grains move through the gas, dust-gas collisions tend to smooth out their motion till they are all moving in similar circumsolar paths with small velocities relative to each other. They will of course tend to collide, but the low relative velocities mean that the collisions are not catastrophic; if the encounters are gentle enough, the particles will stick together rather than breaking each other apart.

And here is the first question. How gentle is "gentle enough"? Today, for example, asteroids in their orbits are continually colliding with each other and with interplanetary dust; they are not, however, building up to form a new planet. On the contrary, they are self-destructing: the collisions tend to splatter material away, slowly or catastorphically eroding the asteroids depending on the size and velocity of the material they encounter. This is almost certainly the reason that stone meteorites show cosmic ray ages less than fifty million years: the parent bodies of the meteorites have existed for four and a half billion years, along with the rest of the solar system, but the cosmic ray "clock" is turned on only when some collision reduces them to roughly meter-size (see chapter 22). Once they are that small, further collisions with dust will continually whittle them away, while collisions with larger bodies may destroy them in one go; in any event, by the time something like fifty million years have gone by they have almost certainly been destroyed. We get to see a meteorite only if one of these objects hits the earth before this erosion-destruction time has reached its inevitable limit.

So, then, how do these dust particles ever grow instead of disintegrating upon collision? First, the dense gas of the solar nebula is important. "Dense" is of course a relative term: the gas pressure is on the order of a hundredth to a millionth of a bar (one "bar" is atmospheric pressure today on the earth, 14.7 pounds per square inch), but this is sufficient to slow down and regulate the dust particulate motion so that instead of smashing wildly into each other they bump gently. Second, there must be some "sticking" mechanism—and here again imaginations run rampant.

Iron is a magnetic material, and it has been suggested that small grains will thereby attract themselves to each other. Iron is also a soft, ductile material compared to silicate rock, and so perhaps such small colliding iron grains will weld themselves together. As these processes continue the iron grains grow in size until they can begin to effectively attract silicate grains by gravity; depending upon when such gravitational attraction begins to be important, we might be on the way to providing iron cores for the planets right at the start of planetary accretion. It has also been suggested, however, that such preferential treatment for iron is not necessary. For example, the surface of the moon consists of silicate rocks that have been continually shredded by meteorite bombardment, so that when we went there we found the ground covered with lunar "soil," small particles that were formed by the continual "gardening" of this repeated meteorite smashing. These small particles gained an electrostatic charge through a different sort of bombardment, by charged ions in the cosmic rays and solar wind, and this electrostatic charge made them stick together in clumps. In the early days of solar system formation there is reason to believe such charged ion bombardment would have been even greater, and so the dust grains in the nebular disk must have picked up electrostatic charges, and these would have helped them stick together.

Some combination of such factors was undoubtedly operative, but in fact may not even have been necessary. There was at the time sufficient gas density to slow and regulate the dust velocities by gas drag, so that they were not very different from each other; the dust particles would have had rough surfaces, coated with gas molecules which could partially absorb whatever collisional energies were generated, and this combination of low mutual velocities, gas molecule "cushioning," and rough surface structures which could hook on to each other was probably sufficient to ensure overall accretion rather than destruction as the particles bumped into each other during their circumsolar rotations. (Today the original nebular gas has been nearly totally dissipated, so collisions between asteroids and dust take place at high velocity without any cushioning, and so the effect is destructive rather than accretionary.)

Beyond this rough beginning the accretionary theory subdivides into several distinct models. In one the dust grains simply continue to grow until the planets are formed, but there is evidence in the meteorites that a distinct stage was reached where chemical processing took place in larger bodies, and so most proponents of this theory go along with the "planetesimal" approach, in which bodies with diameters of tens to hundreds of kilometers were formed and had their own definite if transitory existence before they in turn coalesced into the planets. (It was at one time suggested that our moon is one of these planetesimals that somehow survived and was captured by the earth.) In the inner parts of the nebula, where the accreting material was heated by the luminous protosun, only refractory materials (silicates and iron) were solid; hydrogen and helium remained in the gas phase and were not part of the accretionary process. Further out, however, where the temperatures were low, ices of water, ammonia, and methane were available as dust grains; with all this additional mass larger solid bodies could be formed. When these got big enough, they would have sufficient gravity to pull in the hydrogen and helium gas through which they moved. In this manner the inner planets would be small and rocky (the terrestrial planets) and the outer ones would be giant and less dense (the major planets).

Both classes of theories, the giant protoplanets and the small planetesimals, account for the gross differences between the inner terrestrial and the outer major planets. Attempts to test them are now focusing on the precise chemistries of the planets. For example, Ray Reynolds of NASA and Morris Podolak of Tel-Aviv University have pointed out that the two theories predict different ratios of ice to rock in Uranus and Neptune. In the planetesimal scenario, these planets should be composed of rocky cores surrounded by ices, topped off with hydrogen and helium; and the relative abundances of these components should be identical to that of the solar nebula from which they formed. In the giant protoplanet picture, these objects formed by gravitational instabilities, and the final event in their formation was a massive gravitational collapse in which the inner regions must have been heated sufficiently to blow off the outer parts of gas; their composition today would be different from that of the solar nebula. In principle, then, an experimental test can differentiate between the two theories. In practice, unfortunately, we don't yet know the compositions of these planets precisely enough to make the test definitive; either model can be made to fit the data.

A consensus, however, does exist that the planetesimal models can more simply and directly account for what we know of at least the terrestrial planets. The protoplanet theory may yet work out for the outer, more massive planets; and combining the two theories to account for the two types of planet has certain advantages. For example, we have noted earlier (chapter 11) that one particular problem is the small mass of Mars relative to Venus and Earth: why should the latter two be ten times as massive as the former? An answer is possible if Jupiter (and the outer planets) formed by gravitational collapse, which would have taken place before the slower process of dust accretion could have formed the inner (terrestrial) planets. Then the tremendous mass of Jupiter could have gravitationally depleted the dust region in which Mars was attempting to grow, thus stunting its growth. Earth and Venus would have been too far away from Jupiter for such an effect to be felt.

Let us accept then, that while the Jovian planets may have formed by collapse, the Earth and the other inner planets are well explained, at least in principle, by the gradual accretion of dust grains into planetesimals followed by accumulation of the planetesimals into the final planets. Since it's the birth of the Earth we're interested in, we'll follow this process in more detail, but first we should take a closer look at what one planet really looks like—this Earth that we live on—as strangers in a strange land. A very strange land, more unknowable to us than we commonly think, because of its inaccessibility.

It is perhaps momentarily off-putting to think of the Earth as inaccessible, since we are *here,* but in fact it is so. We live on the two-dimensional surface of this three-dimensional sphere, and that third dimension is almost completely barred to us. If we think of the Earth as an apple (bear with me), the deepest hole we have ever drilled has not gone through the skin. How then can we know what the inside of the Earth, the great mass of the Earth, is like? How can we compare it to the predictions of our models until we do know?

SOURCES: Cameron 1985; Podolak and Reynolds 1984

CHAPTER NINETEEN
•
INSIDE THE EARTH

THERE ARE two extreme types of experiments that go with two extreme types of experimentalists. One type of experiment is extremely sophisticated, carried out in air-conditioned, dust-filtered laboratories with computerized equipment pushed to the limits of the state of the art, based on lines of reasoning no less erudite than obscure, explicable only in terms of fourth-order partial differential equations and precise to 0.0001 percent; the experimentalists involved are anal-compulsive. The other type of experiment is of such a low level of sophistication as to be immediately understandable to professors of English, clear and basic, a straightforward order-of-magnitude determination of a simple number. The experimentalists suck on unlit pipes and discuss Kafka.

In the real world such extremes do not often exist; most of us are somewhere in between the two endpoints, fluctuating a bit from one to the other with the passage of time. There are, however, those who take great pleasure in working for years on an instrument itself, tuning it and refining it, cooing over it and worrying over it; and others who impatiently solder and tinker and care only about getting the damn data. I personally take great pleasure in reading about the latter, about those who took a simple idea and a simple measurement and who came up with a profound knowledge. Such is the story of the inside of the earth, which begins with measurements of the Earth's radius and mass.

We have discussed earlier the first Greek and Egyptian measurements of the curvature of the earth, from which its radius follows. The concept by which its mass was determined is equally simple, once Newton's gravity was understood. One simply takes one large ball and one small one and measures the force with which the smaller is attracted to the larger:

$$F = GmM/d^2$$

where m is the smaller mass, M the larger, and d the distance between them. The masses m and M can be separately measured by observing their accelerations when moved by a constant force:

$$F = m \times a = M \times A$$

This combination of measurements gives the gravitational constant G, and then one simply measures the weight of any body m, the weight being the gravitational force exerted upon it by the earth with mass E:

$$W = GmE/d^2$$

where the distance is the distance from the center of the spherical mass to the center of the Earth (essentially the radius of the Earth, neglecting such minor details as the height of the weight above the surface and the oblateness of the earth).

The experiment was carried out in 1797 by Henry Cavendish. The only difficult part was measuring the small attractive force between two bodies of mass small enough to be manageable, and he did this by observing the twisting imparted to a thin wire which held them. The result was extremely accurate—at least it was accurate enough so that no better estimate was made for more than a hundred years and, more importantly, accurate enough to distinguish the true density of the earth from that of our surface rocks.

The rocks which make up the surface of the earth are distinctly variable. They are igneous, solidifying from a melt, or sedimentary, compacted from deposits of sediments, or metamorphic, changed by heat or pressure from either of the other two. Within each group there are a number of chemical and mineralogic varieties, but practically speaking, all of them have the same densities; just about 3.5 grams per cubic centimeter. The surface of the earth is also abundantly covered with water, which has a density of 1.0. If these two components are the major components of the mass of the earth, its density must be between 3.5 and 1.0; if all the water is on the surface and the interior is simply rock, the earth's density must be essentially 3.5, since the surface water constitutes such a small proportion of the earth's total mass. If the interior rock is subjected to pressures intense enough to compress it, the density will be slightly greater; the effect is easily calculable and leads to densities on the order of 3.8.

But the mass of the earth as determined by Cavendish and slightly modified by more modern determinations is 6 x 10^{27} grams, its radius is 6,370 kilometers, and its density is therefore 5.5 gm/cc.

Two simple measurements, an even simpler calculation, and suddenly the earth is not what it seems to be. We cannot be sitting on the surface of a homogeneous ball—something inside the earth is denser than the surface rocks.

People knew immediately what it must be. In 1600 William Gylberde (or Gilbert, as it is currently modernized), physician to the Queen and foremost scientist of Elizabethan England, an emminent astronomer who argued that the fixed stars are further away than the planets and are not fixed in an equidistant sphere at all, a natural philospher who was the first man of substance to support the Copernican view (which in anti-Catholic England came to be regarded as a sporting proposition rather than a subject for burning), published his treatise *De Magnete* in which he concluded that the earth was a giant magnet. This work, one of the first in England to adhere strictly to what we now call the scientific method, detailed many years' experiments on magnetic interactions and concluded that the deflection of the magnet—which as the sea-faring compass enabled Elizabeth's mariners to conquer the oceans—was the result of a natural magnetic field anchored in the Earth as if it were a huge spherical lodestone. From what was already known about magnets (but not yet understood: there is a difference) he proposed that the earth must contain a good deal of iron.

Gilbert's work directed attention to geomagnetism, and in the century that followed it was found that the geomagnetic field was not stationary but drifted slowly westward, as if the spinning earth were holding it not quite tightly enough. This was impossible for a completely rigid magnet, and in 1691 Edmund Halley (of cometary fame, whom we discussed earlier in connection with gravity and Newton, and who, incidentally, was described as "a drunken sea captain" by the supporter of a rival applicant for a university professorship in 1691) suggested that the density and magnetism data indicated an iron core, and the drift of magnetism indicated that the intervening earth region must be fluid, so that the core slipped as it rotated and didn't quite keep up with the earth. This fluid (i.e., molten) earth idea found increased support when the first ideas for the origin of the earth suggested a hot, molten birth out of the sun, and measurements of terrestrial temperatures in the nineteenth century fell into line, showing increased temperatures with depth.

But then somehow the next series of measurements and theoretical calculations took a wrong step. It's an interesting example of the secular fallibility of the scientific method, for none of the people involved fell into the recidivistic errors of religion or mysticism; they measured things and calculated things, interpolated and extrapolated, and simply came up with the wrong answer. It's also an interesting example of faith: faith in the scientific method, which enabled a succession of workers to work their collective way eventually through the confusing morass of false interpretations and to arrive finally at the truth. (And how do we know it's true? How do we know that the next generation may not find today's truth tomorrow's fallacy? The only reply is that we must retreat behind the barrier of "beyond a reasonable doubt." Too many data and too much theory today line up solidly to form a reasonable story: it would indeed take a leap of faith to disbelieve it. So there's no alternative to acceptance. For fun, of course, we may keep that little residue of doubt alive and squirming.)

The early nineteenth-century concept of a thin stable crust floating on a fluid interior seemed totally wrong to such emminent "modern" nineteenth-century physicists as Andre-Marie Ampere in France and Lord Kelvin in England. They pointed to the lunar tides which cause the ocean to rise and recede, and showed that an intervening crust would not diminish just such an effect on the postulated interior earth fluid. As the interior rock rose and fell, it would necessarily crack the solid but brittle crust which, being solid and therefore more dense, would then sink down to the center of the earth like a foundered battlecruiser. Such repeated cyclings would have the effect of bringing the cooler crust continually down to the hot core and allowing the hot fluid to bubble up to the surface, where it would cool, solidify, break up in its turn and sink; the result over the tens of millions of years of history that Kelvin allowed would be a slow cooling of the entire earth, so that by today it must be completely solid. The dilemma about the westward drift of the magnetic field was one that would have to be dealt with otherwise; it was not important enough to overturn the necessities of heat flow in a cooling earth.

This is where we sometimes unavoidably go wrong in science. When conflicting ideas confront each other and we have no firm data on which to base a decision, we have no recourse but to decide according to our "scientific intuition" which idea is the most basic, the most important, the most unalterable, and then to sacrifice the other. But scientific intuition is no more reliable than a mother's love for her murderer/rapist son, few of whom have been jailed without a sobbing woman outside the gates swearing that he is a "good boy" if only the truth were known. Unfortunately the truth is best known through evidence, not intuition, and when we are forced to use intuition we would do well to remember the sobbing women and Lord Kelvin, and go back into the lab to gather some hard dispassionate evidence.

More data, then, were needed; and they were available, coming from the laboratory of Thomas Andrews in Ireland, a chemist unconcerned with the earth but intensely interested in the properties of gases. He showed that there lies a point in the fortunes of any gas which may not be exceeded without wonderful consequences. At temperatures above this point, the "critical point," all differentiation between the concepts of liquid and gas are lost, the matter flowing continuously and without restraint between these two forms. To put it somewhat differently, above the critical temperature no gas can be condensed (defining "condensation" as a precipitous decrease in volume), no matter how high the pressure. Thus all substances whose critical temperatures lie below room temperature (roughly 20° C) are called *permanent gases,* since it is impossible to liquefy them at "ordinary" temperatures by simply increasing the pressure. Since the temperature inside the earth was calculated to lie above the critical point of both stone and iron, it followed that no matter what the earth was composed of, iron or stone, it must be a gas inside. The pressure of the gas would keep the crust forever floating, even if it broke up. The magnetic field might be explained if the gas were iron; the density of gaseous iron was not explicitly known, since we couldn't raise any material in our laboratories to the temperature and pressure involved, but intuition said that when we learned how to, we would find results in accord with the measured density of the earth.

We didn't, of course.

Instead we discovered earthquakes.

Or rather, we discovered how to use them to understand the interior of the earth through which they send signals. The discovery was not an instantaneous one, but certainly had its beginnings in the work of R. D. Oldham, who showed that the reverberations of an earthquake, similar to those of a struck gong or of Noel Coward's women, result in shock waves that are of two kinds: the P waves are compressional waves and travel through any material, while the shear S waves cannot travel through a fluid. In his original work he found that the S waves went through the center of the earth at a much slower speed than elsewhere; he concluded that the center of the earth consisted of a core which was different in some way from its surroundings.

Other geophysicists, analyzing the same data, found that the S waves were not coming through the core at all, but this conclusion was not acceptable until Harold Jeffries in 1926 gave a theoretical argument showing that the rigidity of the earth taken as a whole, measured by surface tidal data, was less than the rigidity necessitated by measured velocities of propagation of the earthquake waves through what was by then called the mantle of the earth. To reconcile the differences, he demanded that the core be fluid. Reevaluation of the S wave data then convinced everyone that they were actually not seen traversing the core, and that therefore the core was indeed a fluid.

By this time meteorites were being understood as samples of the cosmic soup from which the earth must have formed, and the iron abundance in the meteorites was on the order of 30 percent while in terrestrial crustal rocks it is only about 10 percent. We had, then, the following arguments:

1. The earth has a magnetic field, and therefore must contain a large supply of iron somewhere inside it.

2. The magnetic field drifts westward, therefore the iron supply spins laggardly compared to the overall spin of the earth.

3. The density of the total earth is greater than that of the crustal rocks.

4. The terrestrial iron content should be 30 percent, but the crustal rocks show only 10 percent.

5. S waves do not travel through the central region, which is therefore fluid.

Putting all these together, the only conclusion is that the earth has an iron, fluid core. Quantitative measurements of the magnetic field strength, the chemistry of the earth, and its density, all agree that the core is about 3,400 kilometers in radius. (Another seismic wave discontinuity probably indicates a solid inner core, but here we once again reach the limits of the knowable.)

Differences should exist among the terrestrial planets which are dependent on their varying distances from the sun, but the earth is presumably typical of this group of bodies; we will not go into the somewhat subtle differences which exist among the group. However, we should mention one last point before going back to investigate the different theories of formation of the terrestrial planets: the terrestrial moon, or rather the moons, or rather the lack thereof. We will do more than merely mention this: we will admit defeat.

Jupiter has 16 moons, at last count, and Saturn has 21. Uranus has 15 and Neptune 2, but the latter planet is so far away there almost certainly are other small moons out there that haven't been seen yet. The terrestrial planets, by contrast, are practically naked, moon-wise. Mercury and Venus have none, Earth has the one, and Mars has two. But the Martian moons are tiny, puny things, most likely captured asteroids rather than true moons ("true" being used in the sense of satellites born in some way from or out of or in conjunction with the planet around which they revolve. It is likely that the Martian moons are asteroids perturbed out of orbit by a close passage of Jupiter and subsequently captured by Mars, in much the same manner that numerous asteroids have been perturbed by Jupiter into Mars-crossing orbits and then swung by Mars into earth-crossing orbits, whence the meteorites.)

Which leaves our moon as a unique object, and so it is. It is the only moon in the solar system with dimensions approximating its planet; but whether or not it is a true moon in the sense outlined above is still not known. It may have been formed in some manner out of or in conjunction with the earth, but none of the theories describing this process are satis-factory, as mentioned previously. So we shall pretend either that the moon does not exist or that it formed, like the Martian moons, in some manner unconnected with earth formation and so is extrinsic to that question. It should be noted, of course, that the moons of the major planets are not so inconsequential: there are too many of them to be so lightly dismissed, and theories of origin of the Jovian planets must include formation of regular satellite systems. A mimicry of solar system formation is clearly indicated, perhaps along the lines of the Cameron models, but just as clearly the terrestrial planets are different in this respect and so perhaps two different types of planetary formation process should be invoked. At any rate, we are now concerned with the origin of the earth, and so shall proceed to a consideration of the most likely ways to make the terrestrial planets *sans* moons.

SOURCES: Brown 1974; Brush 1980, 1982

CHAPTER TWENTY

•

ACCRETION AND DIFFERENTIATION

THE SIMPLEST WAY to accrete a solar nebula into planets would be for the dust grains simply to stick together and grow like Topsy. And while Occam's Razor reminds us to keep our theories simple, we must also remember that the world was not created solely for us—neither as a Garden of Eden to inhabit nor as a straightforward puzzle to charm away a pleasant hour. Things insist on getting more complex.

As the nebular disk formed, gravitational energy was released and both pressure and temperature increased. At first the heat was radiated away, but when the disk became optically opaque it began to trap the heat. Obviously this trapping was a function of density, and so both the heat and pressure build-ups became greatest at the center, decreasing further out. Typical maximum calculated values are 0.001 bar pressure and $\sim 2000°$ K, which would have been sufficient to process the dust chemically and mineralogically. Once this was realized, it opened the way to a new line of investigation. No longer was the field restricted to astronomers and physicists; now the chemists began to get into the act, and the field of cosmochemistry was born.

Still trying to keep things simple, the first attempts to unravel the chemical processes began with the assumption that the nebula itself began as a chemically homogeneous mixture. This made sense, given the ideas of how it formed, whirling and turbulent and mixed by magnetic flow. It made sense also based on the observations we had at the time. Chemically we always knew things are different in different parts of the solar system today, but that could be accounted for by chemical processing proceeding through time; and isotopically things looked to be exactly the same everywhere.

This was the important clue. Isotopes are identical atoms, except for differing numbers of neutrons in their nuclei. Their extranuclear electronic configurations, in particular, are identical. Since in chemical reactions the parts of the atoms that react are the extranuclear electrons, such nearly identical atoms are precisely identical chemically. If the original solar nebula was isotopically homogeneous, then no matter what chemical processing went on it would remain so.

Thus an original nebula that was a well-mixed interstellar cloud, therefore both chemically and isotopically homogeneous, would after a serious bout of chemical processing remain isotopically homogeneous even though chemically heterogeneous—which is just what we saw up through the 1960s. Except for effects which have clearly taken place after the formation of the planets, such as radioactive decay and nuclear reactions induced by cosmic rays and mass fractionation, isotopic ratios were identical everywhere on the earth, in all the different classes of meteorites, and on the moon. Theoretically there was always the possibility that solid dust grains might have formed early in disk history, before it was well mixed and homogenized, trapping and preserving isotopic differences in their interiors, but the nebular temperatures were thought to be sufficient to vaporize any such early grains, and so their material would be mixed back into the protosolar/disk reservoir before the planets began to form.

Today we know this picture is too simple. As experimental techniques in the 1970s and 1980s became more sophisticated, sensitive, and precise, particularly at the University of Chicago and CalTech, isotopic differences were in fact found. But it's easier to begin talking about a mythically homogeneous nebula and then introduce such complications later, rather than dive right into the morass of variations.

Picture a nebula, then, that condenses and flattens into a disk with a strong temperature and pressure gradient. Dust grains form and chemically react with the gas in which they are floating, bump into each other, aggregate, and finally seal themselves off from further reactions by burying themselves under other grains. Some of this dust in the inner regions may fall into the sun, some at the outer edges may spiral off into space or be left behind out there, still whirling around unseen and unknown; some will form into planets, and some will make it only part way. This latter group is presumably the asteroids, which thereby represent a valuable checkpoint to our theories—made more valuable still by their tendency to get thrown out of stable orbits into paths which cross the earth, and thus to fall to earth and to our laboratories as meteorites. Some of these meteorites, the basaltic achondrites, are similar to volcanic rocks on earth and can be mineralogically recognized as the products of melting. Therefore they must have formed on bodies much larger than the present-day meteorites. (The ability to hold internal heat is a function of the body's mass-to-surface ratio, since heat is lost only at the surface. The mass increases with the volume, which goes as the cube of the radius; the surface area increases with the square of the radius. Therefore as the radius increases, so does the mass-to-surface ratio, and the ability to hold sufficient heat to induce melting.) Mineralogical/petrological studies on various meteorite classes give indications of the rates at which they cooled, the slowest of which vary from 1 to 10 degrees per million years, indicating parent bodies of tens to hundreds of kilometers in radius, about the size of the largest asteroids. (Recent work indicating faster cooling rates has been interpreted as due either to smaller parent bodies or to the possibility that the meteorites observed may have been formed near the surface of their parent bodies.)

Other meteorite classes, particularly the carbonaceous chondrites, and very particularly one subset called group CI, contain water and carbonaceous compounds which are destroyed by rather minute quantities of heat, and so bear in their constitutions evidence that they have never been exposed to high temperatures: they have never been chemically processed like most of the dust grains in the nebula and like the mass of the earth after it formed, so they represent evidence of the most primitive matter out of which the present solar system formed. At least that's the concept, and it seems to bear a strong resemblance to reality. It is in minerals of these meteorites, for example, that the largest isotopic differences are found. They are unique resources, then, for even the smallest of planets is large enough to undergo melting and a calamitous variety of chemical and mineralogical processing induced by the generation of radioactive heat and the trapping of such heat by its sheer mass; the meteorites, or at least some classes of meteorite, have avoided this and therefore offer us an untouched Rosetta Stone bearing the record of what came before it all began. Unfortunately, of course, the language of that Rosetta Stone is not modern-day English: it is chemistry and mineralogy.

It was 1802, when the true origin of meteorites as stones from the sky was not yet recognized by most people (like Thomas Jefferson), that the first substantive chemical study was undertaken in England by Edward Howard and Jacques Louis de Bournon. They studied four rather ordinary-looking rocks that had been seen (or claimed to have been seen) to fall out of the sky. The only thing mildly unusual about them was that they were sprinkled through with small globular grayish inclusions, ranging in size from "a small pin's head to that of a pea," and that they contained small particles of metallic iron, which on earth is nearly always found not in this state but rather oxidized to some combination of FeO or Fe_2O_3. Further analyses showed that these metallic grains consisted of iron alloyed with nickel, and these, together with the small globules, were and are today enough to make these meteorites recognizably different from all terrestrial rocks. The globules were named "chondrules," from the Greek *chondros,* which means a small seed, and the meteorites which contain them are called chondrites. Everything which contains chondrules is a chondrite, but not all meteorites do. Other important meteorite classes recognized today are the irons, which are pure (or nearly so) nickel-iron, the achondrites, which are stones without chondrules, most of which were produced by igneous processes somewhere out there, and the stony-irons, of obvious composition.

Mineralogical studies of the chondrules show that they were once molten drops of silicate rock, but they are intimately mixed with other minerals which rapidly decompose at elevated temperatures, so they could not have been formed *in situ.* Later chemical studies in the 1930s by the German husband-and-wife team of Walter and Ida Noddack showed that the chondrites are mineralogical mixtures of chemical elements with quite different characteristics. A generalized grouping of the geochemical behavior of the elements can be thought of, with all the elements organized into one of three major categories: the *lithophile* elements are those which form silicate minerals, the *chalcophile* elements form sulfide minerals, and the *siderophile* elements dissolve themselves in metallic iron and nickel.

The natural processes that have operated on earth have efficiently separated these groups, with the siderophile elements sinking down with iron to the core, the lithophile elements forming our rocky crust, and the chalcophiles falling into regional ore deposits. But in the chondrites all three groups are present at reasonable levels of abundance, indicating that these meteorites have not been chemically processed as earth rocks have. Add to this the mineralogical evidence of igneous chondrules existing side by side with low temperature minerals, together with radiogenic ages of four and a half billion years, and you have a picture of an aggregate of material that has not been changed or chemically altered since it formed at the very beginning of the solar system.

The chondrites in particular seem to be a good average of solar nonvolatile material, and therefore a good starting point for calculations of the aggregation of dust into planetesimals and on into planets. The chondrites themselves can be subdivided into several classes, and arguments as to which of these classes is the most primitive are still continuing. The consensus is that carbonaceous chondrites of class 1 (C1) best satisfy the criteria; this is a detail of no small concern to active workers, but need not concern us here.

The establishment of some class of chondrites as representative chunks of the solar nebula was the beginning of our chemical understanding of solar system history, and it was due to the insight and intuition of one man, Harold C. Urey (photo 20.1). Before Urey began to preach this idea, the evolutionary history of our world began with the geological evidence, very little of which extends back beyond a half billion years—nowhere near the beginning. The astronomers were arguing about theoretical models and constructs, and they could have gone on arguing till Doomsday without getting very far, because a theoretical argument is incapable of proving anything or of itself being proven true or false unless and until you can get your hands on something that you can carry into the laboratory and take apart or dissolve or put under a microscope. And until Urey told us, lectured us, preached at us, and convinced us, nobody knew what to take into the lab or had confidence in what the results might mean.

His message was simple—chondrites are the unchanged relics of the primeval matter from which the earth emerged—but not obviously true.

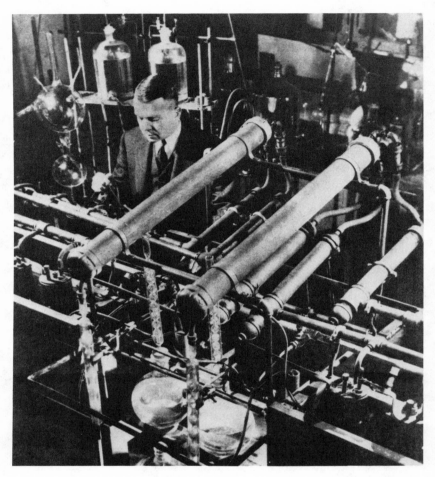

Photo 20.1. Harold C. Urey, the first modern cosmochemist, in his laboratory, ca. 1930. MRS. H. C. UREY

His argument was made forcefully when he and Hans Suess, a colleague at the University of Chicago, published in 1956 a paper reviewing all the then-known chemical analyses of chondrites and showing that they were all closely identical, close to solar analyses in all except the volatile elements such as the rare gases, and close to what would be expected in the solar nebula. I remember reading that paper in 1960, when I had just joined Cornell University as an assistant professor, and when I took the trouble to check into the Suess-Urey claims I seriously wondered at the Suess-Urey *chutzpah*.

The point I picked on was the abundance of aluminum, which Suess and Urey stated was 1.3 percent in all chondrites and therefore also in the solar nebula. What they actually said was that aluminum's abundance was 1.3 percent in all "preferred" analyses. That word "preferred" struck a suspicious-sounding note, and I went back to the original published papers to check the analyses. I found, in fact, that there were quite a few chemical analyses that gave largely discrepant aluminum values, some as low as 0 percent, some as high as 7 percent.

Now aluminum, as was well known, is a bitch to measure chemically. It precipitates as a flocculent, messy hydroxide which easily scavenges other metals and carries them along, and is itself readily adsorbed in other procedures; it would be easy to get erroneous values for its abundance in a matrix as complex as the chondrites. On the other hand, the analyses which gave values of 1.3 percent were just as likely to be in error as those which gave different results. Why were those values "preferred"? I asked Urey, and he replied because those were the analyses that gave the right answer—in other words, he had found nothing in the chemical procedures to set those analyses apart, nothing in the reputation of the workers who performed them, nothing, in fact, except that they gave the value he wanted. If the measurements ranging from 0 percent to 7 percent were valid, his idea that the chondrites were chemically identical was not valid— and that idea was necessarily central to the conclusion that they represented the undifferentiated solar nebula. But the reasoning was circular: the chondrites were chemically identical because the data said they were, but only if one threw out the data that said they weren't!

Terribly unscientific. At the time I had at my disposal a nuclear reactor, and with it a new means of chemical analysis. If a sample of aluminum was irradiated with neutrons, it became radioactive and threw out a detectable gamma ray of unique energy, thereby becoming reliably analyzable without going through any difficult chemical precipitations. I wrote to museums all over the world and received a supply of chondrites, including some where previous analyses had shown 1.3 percent aluminum and others with widely variant results. I tested the method by dropping into the reactor the "normal" chondrites, then taking them out and measuring their gamma ray emissions: in all cases the results showed 1.3 percent aluminum. When I was satisfied that I knew what I was doing, I dropped in the Alfianello chondrite, which had been previously analyzed twice and had shown 0 percent in one analysis and 0.47 percent in the other. I had the answer within a few minutes: 1.3 percent aluminum.

I still remember the shock. I couldn't believe it. I then measured the Melrose chondrite, which at 2.6 percent had the highest previous analysis I had been able to locate. The value I measured was 1.3 percent. Over and over that long day and into the night I measured one meteorite after the other, and they were all 1.3 percent (within a tight experimental error). I called Professor Urey the next day, and I swear I could hear him smiling over the phone; he was pleased, but not at all surprised. I'm still not sure if it was insight, intuition, or good luck, but he was certainly right.

He wasn't right in everything. He was wrong, in fact, in practically every detail he argued for: for example, he asserted as forcefully as only he could that the moon was an undifferentiated original condensation from the protosolar nebula and therefore would, as soon as we managed to get there, tell us everything we wanted to know about the origin of the solar system. This argument of his, in fact, was the original and strongest scientific argument for putting our men on the moon. And if the final decision to race the Soviets there was a military and political one, Urey's argument and personality were instrumental in turning what could have been a disgraceful waste of money into perhaps the greatest scientific voyage in history. But now we have gone to the moon and come home again, and the rewards Urey promised us have not been seen. We still don't even know where the moon itself came from; we know only that he was wrong, that the moon is a later creation and does not bear in its innards the tale of creation.

Never mind. He was wrong in one detail after another, but he was right in his concepts, in an overwhelming vision of that creation. He was the first person to tell us that chemistry—everyday, terrestrial chemistry—was the path to the heavens. He singlehandedly created the field of cosmochemistry—the "geochemistry" of everything we can get our hands on, the earth and the moon and the meteorites and whatever else we can reach out for and grab—and convinced us that there was a great store of truth hidden in those stones which, when combined with the theories of the astronomers, could unlock the mysteries of the universe for us.

And he was right. And he fathered a generation of cosmochemists in Chicago and grandfathered another generation in California (which is why so many names mentioned here are associated geographically with those regions) and sired bastards like me all over the world. And he stimulated and encouraged and provoked all of us, and everything we know today about the origin of the earth is a monument to his stubborness, personality, insight, intuition, and luck.

He was a gruff man, impatient and demanding; and somehow—for his honesty, perhaps, or his genuine passion for knowledge and understanding, or perhaps simply his interest in us—we all loved him. God rest.

During the next two decades after his sermon that the chondrites were the Word, many workers using the technique I had used, "activation analysis," and other newly developed analytical methods, removed one by one the old discrepancies between what the older analyses showed and what Urey "knew" the actual chemical abundances must be. Some analyses were more intractable than others. Iron, for example, continued well into the 1960s as a prime stumbling block; it is one of the most abundant nonvolatile elements and couldn't be ignored, but its measured value in the chondrites didn't match its abundance in the sun. For a decade meteoriticists tried to improve their techniques, and then in 1969 the solar spectroscopists found that it was they who had been wrong: their new value agreed perfectly with the meteoritical abundance.

Combining the solar abundances for the volatile elements which have been lost from the chondrites with the nonvolatile abundances measured in the chondrites, a "cosmic abundance table" can be put together. The name is a misnomer, since it refers to the abundances in the primitive solar nebula and not to those in the entire cosmos, or universe, but it's one of those mistakes by now well established in tradition, and is undoubtedly permanent. Once we know the chemical abundances in the nebula, we can begin to make quantitative calculations as to what happened as the dust grains condensed from the gas under probable conditions of temperature and pressure. Lawrence Grossman of the University of Chicago, a pioneer in this research, offers the condensation of calcium as an illustrative example.

The abundances of the pertinent elements in C1 chondrites (the "cosmic" abundances) are listed below.

Element	Abundance
SiO_2	17 percent
Al_2O_3	3.4 percent
CaO	1.3 percent
TiO_2	0.06 percent
MgO	15 percent

Note that the data are listed in terms not of the elements, but of their oxides. This doesn't make sense, of course, since they are present not in the form of simple oxides but as complex molecular mixtures such as, for example, *amphibole:* $Ca_2Mg_5(SiAl)_8O_{22}(OH)_2$. They are listed this way because it is *tradition.* Today geochemical analyses are made by pushing a button which turns on a minicomputer which directs an atomic absorption spectrophotometer to whirr and click and collect data and store it on a floppy disc while a tan-legged (I'm writing this in Miami) graduate student writes to her boyfriend in law school. Today all the romance has gone out of the profession; but in the good old days chemical analyses were made by dissolving things in beakers—and silicate rocks are nearly impossible to dissolve, necessitating reagents like hydrochloric and nitric and even hydrofluoric acids, and heating to boiling, and stirring and stirring and stirring, which is why all old geochemists have scarred and burnt fingers, and no good clothing at all. One can't help thinking of Madame Curie, like one of Macbeth's strange ladies stirring up her old iron cauldrons of dissolving radioactive rocks with a monstrous big stick, and contrasting that vision with the giggling young graduate students of today, smoking their Virginia Slims while the computer stacks and sorts, regresses and regurgitates—both of them, Madame Curie breathing radioactivity and the young girl smoking her cigarette, doomed to die of cancer—and thinking, you've come a long way, baby, but not far enough. Oh well. Nostalgia is such sweet sorrow, and today the romance is gone from the old profession, all gone—and, it has to be admitted, a good deal of the drudgery and inaccuracy is gone, too. But in those good old days the final analysis was made by weighing a precipitate which was usually composed of the simple oxide, and so today we remember those days by still listing chemical analyses as oxides. It would make more sense, of course, to list the elements themselves, e.g., Ca, Si, Mg, etc., and more and more frequently these days this is done; but to remember and to honor those great men who could dissolve a stone and bring forth from it the simple oxides

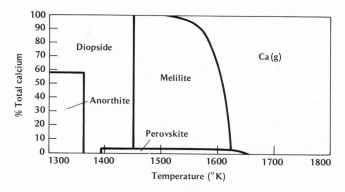

Figure 20.1. The Fate of Calcium as the Nebula Cools   LAWRENCE GROSSMAN

in quantitative exactitude—more miraculous, surely, then striking it to bring forth merely wine—men like H. B. Wiik in Finland and Eugene Jarosewitch in Washington and Oiva Joensuu in Miami, we'll continue to list those beautiful oxides.

Let's start now in the forming solar nebula, at a gas pressure of about 0.001 atmospheres, with a chemical composition as given above, and with a temperature somewhere above 1700° K. At these temperatures all the calcium is an elemental gas, composed simply of individual calcium atoms floating free. This is indicated by the area labeled Ca(g) in figure 20.1.

As the nebular temperature drops off, nothing important happens to calcium until it reaches 1650°, where the calcium titanate mineral perovskite (CaTiO$_3$) becomes stable. As Ca, Ti, and O atoms in the nebular gas collide below this temperature, they spontaneously form this solid mineral. As the temperature decreases these mineral grains continue to grow, until at about 1620° another calcium mineral involving magnesium and silicon begins to form; this is melilite (Ca$_2$(Al,Mg) (Si,Al)$_2$O$_7$). By 1600° all the titanium in the nebular gas has been used up and so no more perovskite can form; all the remaining calcium now goes into melilite. As you can see from the table, calcium is less abundant than aluminum, magnesium, silicon, and oxygen, so its form is not limited by the availabilities of these elements as it was by the less abundant titanium. This, then, would be the end of the story if the gods were indeed merciful, but no: things get more complicated. At 1450° melilite becomes unstable; its atoms rearrange themselves into a more compact structure with no place for aluminum, and so the A1 atoms are thrown out; in the rearrangement the condensate gains Si instead and becomes the new mineral diopside $(CaMg(SiO_3)_2)$, so that now a few percent of the calcium is in perovskite and most of it is in diopside. Then when the temperature dips below 1400° the perovskite becomes unstable, self-destructs, and the liberated calcium atoms are formed into diopside. At 1360° another calcium mineral becomes more stable than diopside; this is anorthite $(CaAl_2Si_2O_8)$, which will suck all the calcium out of the diopside crystal lattice until all the aluminum in the nebula has been used up, and then of course the process must stop. Given the chondritic abundances, the result is about a 55—45 mixture of anorthite-diopside.

This is not actually the ultimate fate of calcium. As the temperature continues to drop, other mineral reactions continue to occur, the end result of which is that most calcium is in a complex form called plagioclase $((CaAl, NaSi)AlSi_2O_8)$, but the path to that mineral is more complex and no more illustrative than the high temperature sequence detailed above. The first point is that the minerals formed as the temperature drops depend not only on the temperature but also on the relative abundances of the chemical elements involved; if more titanium had been available, for example, more perovskite and less melilite would have been formed. The second point is that the minerals, once formed, are not permanent; as the temperature drops they may become unstable toward internal rearrangement with concomitant expulsion of particular atoms, or another mineral may become more stable and take over.

Obviously it would be an extremely complicated calculation to begin not only with the simple set of elements listed above—Ca, Al, Ti, Si and O—but to include all the naturally occurring elements and their mutual interactions and competitions, but this is what cosmochemists are paid for. Such calculations can be done, and in fact only the one or two dozen most abundant elements have to be considered, at least at this stage of the game, to give results which correlate at least qualitatively with reality. In other words, we're on the right track. Some typical results show that grains forming near the sun, at higher temperatures, would be different from those forming further away. And this difference will of course be reflected in the final planets formed at each distance. Assuming that each distance had its own characteristic temperature, the chemical and mineralogic composition of the planet formed at that point can be calculated as if it formed at that particular temperature, which varies inversely with distance from the sun.

The results of the calculations show a progressive decrease with density in the region of the terrestrial planets, due to a higher proportion of less dense minerals forming the final state of the condensation process at lower temperatures; the calculated densities range from a high of 5.4 at the radius of Mercury's orbit to a low of 3.9 at Mars', and these correspond quite well to the observations. Detailed chemical calculations also indicate the various planetary chemistries: Mercury, for example, should have all its iron in the metallic state, Venus and Earth progressively less metallic iron and more chemically oxidized, while Mars should have no metallic iron at all.

The major, Jovian planets are completely different, but the differences (in this scheme) are due to the operation of the same processes. At their great distance from the sun the temperatures are so low that the ices of water, methane, and ammonia can condense to solids. These elements— hydrogen, oxygen, carbon, hydrogen, and nitrogen—are more abundant than those of the silicate minerals, so a regime cool enough to allow them to condense will naturally lead to the formation of much bigger planets; and since these ices are also much less dense than the silicates, the Jovian monsters are themselves much less dense than the silicate planets. Of course the silicates are present out there, too; the low temperatures in no way prevent their formation, but they are "swallowed up" by the more abundant ices.

This first model of planetesimal formation, called *homogeneous accretion,* had great success. Clearly we were on to something. But, in the great scientific tradition of never letting well-enough alone, people began to probe into details and inevitably found a few that disturbed them.

Water, for example, should form abundantly in the icy regions of Jupiter and Saturn, but not at all where Earth formed. Pure water is not necessary at this stage, for many minerals can become hydrated—they pick up water molecules and incorporate them into their structure, whence they might be liberated at a later stage in Earth history and bubble up to the surface—but the predicted temperature of formation for Earth is too hot even for the condensation of hydrated minerals. Now water, of course, is not a major constituent of Earth, but it is a rather obvious thing not to have around, since without it no life could possibly exist. Not to appear ingrates, we can't ignore its absence from our theory.

There are other chemical problems. The core of the earth is iron: how did all the iron in the little planetesimals congregate together in the final planet to form a central core? Easy enough, the planet must have melted, and the dense iron droplets would then have sunk from everywhere down to the center. But then all the siderophiles should have dissolved themselves into the iron phase and gone with it, and they haven't. We have a large percentage of nickel, for example, in the mantle and crust of the earth. And the form of iron in the earth is another problem. Iron can exist in the metallic state, with an oxidation number of zero, or it can be oxidized to the $+2$ ferrous or $+3$ ferric state, but the two extremes are not compatible with each other, particularly in the presence of water. Metallic iron will react spontaneously with ferric iron and both will change to the intermediate ferrous form. So if the earth melted, bringing into contact all the metallic and ferric iron, how could they have segregated into core and crust without first changing into ferrous?

As with all such problems, there are two general lines of solution. One is to revise the chemical considerations within the confines of the theory, the other is to search for a new theory. And since there is more than one scientist in the world, both these attempts have been carried out simultaneously, and both have borne both fruit and fault.

The adjustment to the theory is called the *heterogeneous accretion* model, worked out in various details by, among others, Ted Ringwood in Australia and Syd Clark and Karl Turekian at Yale. It envisages planetesimals that accrete quickly, keeping pace with condensation from the nebula. In the example we discussed above for calcium, for instance, not all the equilibrium processes would have had a chance to occur. If the particles accrete rapidly enough, the first (high-temperature) minerals become buried as more stuff accretes around it; the inner core therefore is essentially isolated from the outside world. In the case of calcium, a melilite core at 1450° could not transform into diopside because it wouldn't be able to pick up silicate from the nebula; the outer shell of the forming grain would change to diopside, but the inner core would remain melilite. So instead of everything in the nebula going through the entire range of transitions, the inner core of the forming grains would "remember" their formation temperatures and remain faithful to their high-temperature mineralogies.

This is particularly important in the case of iron. In the homogeneous accretion model a planet forming at earth's distance from the sun would be composed of planetesimals having a mixture of metallic, ferrous, and ferric iron. When the core is formed, these should all mix and the problem mentioned above is encountered. In the heterogeneous accretion model, however, the first iron to condense is the high-temperature form, which is metallic. Planetesimals with this metallic iron may begin to accrete before the lower-temperature ferrous and ferric leave the nebular gas, and so the earth as it formed could already have had a core of metallic iron. Later the more oxidized forms might condense onto that metallic core to form the mantle and crust, without ever mixing with the iron in the core (although there are problems with the rapidity of accretion necessary to achieve this).

At the same time, the iron problem was solved within the confines of the homogeneous accretion model by thermodynamic calculations which showed that iron behaves in an unexpected manner at great temperatures and pressures. Under such conditions a factor that is so minor as to be irrelevant at the surface of the earth becomes overwhelmingly important: changes in volume. The minerals formed of ferrous iron, such as the simple oxide FeO, occupy more volume than the combined minerals of metallic iron and ferric iron. Because of this, if FeO is subjected to the high temperatures and pressures characteristic of the inside of the earth, it will spontaneously "disproportionate" into metallic and ferric forms in order to squeeze into the available space. This necessity to occupy the smallest possible volume becomes the dominant factor, taking precedence over all other considerations, when the pressure to squeeze in becomes so tremendous. Thus these seemingly incompatible forms are actually the most compatible of species, under deep earth conditions.

During the 1970s and 1980s different groups have been putting their money and time onto different aspects of the homogeneous or the heterogeneous accretion models, with the interesting result that these two opposite-sounding alternatives are, like communism and capitalism, coming to resemble each other more and more. At times it seems that you can't even recognize the players without a scorecard. Let's take a look.

SOURCES: Suess and Urey 1956; Wood 1977, 1979

CHAPTER TWENTY-ONE

•

MAIDENHEAD REVISITED

THE VIRGIN pre-solar interstellar cloud, impregnated by a high-pressure disturbance, gave birth to the sun and planets through a process of gathering itself together and chemically differentiating into the bodies we see today. Early theories (circa 1960s) thought in terms of homogeneous planets accumulating via accretion of chondrite-like pla-netesimals. In the 1970s heterogeneous accretion theories took a different tack in an attempt to explain today's chemically differentiated planets. In the 1980s these two approaches have merged and variegated into a galli-maufry of approaches, nearly all of which involve at least some degree of heterogeneity and agree in principle while differing in detail. The reason for the variations is that no one model can yet explain satisfactorily all the observations, and different people put different emphasis on different observations. One picks what one thinks is important, explains it, and waves one's hands a bit to dispel what embarrassments remain stinking up the air. A discussion of two or three of these models will indicate the degrees of agreement and disagreement among the lot.

Ted Ringwood (Australia) concentrates his ideas on the formation of the earth and the other terrestrial planets from planetesimals which were essentially Type I carbonaceous chondrite accumulations. During the en-ergy-releasing impacts involved in the accretion process of the primitive earth, whatever oxidized iron was present reacted with the abundant reduced carbon to form metallic iron; the heat involved depleted the forming earth of the volatiles originally incorporated in the carbonaceous planetesimals. This model necessitates a hot accreting earth; the heat must come from the energy of impact as the planetesimals fall into the growing earth, and this means they must fall in rapidly enough to essentially bury previous impacts before their heat can be radiated away from the surface. When a heat-generating impact occurs in a slow accretion process, the heat is lost to space; eventually, after the site has cooled down, another planetesimal arrives, and in this way the planet grows without getting too hot. The Ringwood process insists that the planetesimals arrive too quickly for that: the temperature of the growing planet must be high enough to volatilize and lose the volatiles, necessitating an accretion time for earth of less than one million years.

Both the rate of impacts and the energy liberated per impact increase as the planet grows. At first one small planetesimal collides with another at a velocity small enough to allow them to accrete rather than simply to smash each other to dust; the heat generated is necessarily small. Toward the end of the process we have a nearly full-sized earth sucking in planetesimals via its gravitational field: a planetesimal falls in much as a meteorite does today, and the full energy of impact craters the surface and delivers a significant supply of heat. Thus the beginning planet is a small body in which the temperature does not rise high enough to volatilize the volatiles or to reduce the iron, surrounded finally by higher temperature layers depleted in volatiles and with their iron reduced. The outside volatile-depleted layers are a more dense accumulation, and so the planet is gravitationally unstable and it eventually overturns, the heavy outside stuff sinking in toward the center. The most dense stuff, the metallic iron, rapidly forms a central core. The lightest stuff, the oxidized original material, floats up and mixes with the volatile-depleted silicates to form the mantle. Under these conditions (with a bit of hand-waving) the process takes place so quickly that equilibrium between metallic core and silicate mantle is not achieved, and so all the siderophile elements in the silicate are not sucked into the sinking iron core. Finally the crust and atmosphere are formed by chemical reactions in the mantle whereby the lightest compounds and minerals float up to the surface, the gases bubbling out.

This model is an example of how the homogeneous and heterogeneous concepts can become fused: it begins with the accretion of homogeneous C1 planetesimals, but the growing earth becomes heterogeneous as the later accretions become successively more volatile-depleted and the iron more metallic. It provides a good overall description of the earth today, but there are some specific faults pointed out by Ringwood himself. If the core-mantle formation process proceeded rapidly enough to prevent equilibration, the resulting mantle should not be so well-mixed chemically as it is observed to be today. Furthermore, the siderophile elements in the mantle should be there in their primordial ratios, which should be those in the C1 meteorites. But such ratios as Ni/Ir are far from the same today in the earth's mantle and the C1's. At this point you can make your choice: you can regard these objections as terminal, or you can invoke the four and a half billion years of mantle convection and geochemical processes that have been taking place since these early events to account for today's observations. This is an unavoidable caveat to all reconciliations of earth formation theory with observations: the formation of the earth took place four and a half billion years ago, and the observations all took place essentially this morning.

Keeping that in mind, let's pass on to another model. Ed Anders of Chicago and John Morgan of Australia, Chicago, Virginia, and a half-dozen other points in between, have led a multivariant group at the University of Chicago to put together over the years a system of layered planetary accretion in which the planets are originally onions, formed skin by skin with each skin chemically different because it forms under conditions of different pressure and temperature at different times and/or places in the solar nebula. Their model is good in that it doesn't dodge specificity: it makes use of experimentally determined chemical parameters to model planetary formation with high precision. The model is bad in that it invokes high precision to such a degree that it loses touch with reality. The Chicagoans identify five "well-defined" (i.e., well-known) chemical elements that are characteristic of the five components produced by not-well-known fractionation processes in the nebula; for example they choose uranium as the characteristic nonvolatile lithophile element and potassium as a volatile element. Then from the known abundance of uranium in the earth and the known terrestrial K/U ratio they can infer the total earth abundances of all other elements that share these characteristics.

Beautiful. But for the "known" abundance of uranium in the earth they choose a value of 14.3 ppb (parts per billion), while other people have made estimates which range from 8 to 28 ppb. For the K/U ratio they take the well-known approximate value of 10,000 determined in a wide variety of crustal rocks and refine it somehow to 9440. But all of these data are from *crustal* rocks only, and there are a few terrestrial determinations that differ fairly widely from this value. Samples that may reflect mantle abundances show values as low as 3,000. So when specific values for U and K/U are taken, and specific values for other elements are calculated from these values and then compared to estimates made for other planets, errors can begin to mount up. To my mind the value of the Chicago model is not in its rather doubtful hold on reality but in the challenge thrown out by its specificity—when it says with such blase self-assurance that the abundance of such-and-such in the earth is this or that, it becomes the most wonderful challenge to experimentalists to go out and prove the assertion wrong. It is the fearlessness of the model and the modelers in throwing out calculations of such precision that will prove both its downfall and its usefulness; the more it's proven wrong, the more useful it will have been.

Such an approach can be contrasted with Karl Turekian's more qualitative model in which the high-temperature minerals condense and are accumulated into planetesimals which begin to aggregate into planetary cores, followed by successive layers of materials which aggregated as they condensed at successively lower temperatures. The lowest temperature material, the last to condense, is the C1 aggregation which falls onto an earth nearly completely formed, bringing with it a light powdering of volatiles which will constitute the oceans and atmosphere. In this model the core is formed naturally, and there is no total melting necessary either for its formation or for that of the atmosphere. It is to a great extent the precise opposite of the Ringwood model, and yet it produces the same earth from much the same material, which shows the value of energetic hand-waving when building models. The value of building such models is that they direct future experimental work instead of leaving the experimenters to wander lost and bewildered without points of reference in the strange world of search and discovery.

Even with such disagreements, most of the current models have more in common than in contrast. They are pretty well agreed, for example, that the earth's inventory of volatiles have come in at the end of the accretion process. Most favor C1 meteorites as carriers, but John Lewis of MIT, founder of another school of thought, favors C3 chondrites—with a postulated orbit reaching out near the orbit of Mars—as the vehicle. If it seems as if such differences are minor, so they are. While the various authors quoted can and do defend their theories vigorously and attack the others' vehemently, they are all in agreement about the basic points.

The first models of planetary formation, begun by Harold Urey, envisaged the chondritic meteorites as the building blocks; the planets formed as homogeneous aggregations, and the chemical heterogeneity we see today—for example, in the core, mantle, crust, and atmosphere of the earth—was thought to be the result of chemical differentiation after the planets formed. It is clear now that at least some of this differentiation must have occurred before the formation of the planets, during the processes of condensation from the nebula, as both temperature and pressure varied as a function of radial distance from the sun, vertical distance from the plane of the ecliptic, and time. After all, the differences between the terrestrial planets and the major planets, taken as groups, must be due to just this sort of thing: we cannot account for the large size and chemical differences on the basis of initially identical planets that then chemically disturbed themselves into two such totally distinct aggregations; rather they must have gathered themselves out of different chemicals because of the different ambient conditions in their different orbits. Once this is granted, it must follow that smaller differences in the orbits of the terrestrial planets must also have caused some smaller differences in the material which gathered to form Mercury, Venus, Mars and Earth.

But the original pattern of this chemical fractionation is difficult to unravel from the whole cloth we have today. Consider the CAIs, Calcium-Aluminum-rich Inclusions, which seemed on their first appearance to be a gift from the gods—and so they undoubtedly are, but we forgot to remember that whom the gods want to endow, they first drive mad.

The gift, thrown down straight from the heavens, arrived in 1969 when the United States was driving hard to achieve John Kennedy's goal of landing a man on the moon by the end of the decade. The engineers were putting together the equipment, the military were licking their chops and already devising star wars scenaria, and the scientists were working on techniques and instruments to measure the lunar rocks which would be brought back. Just after midnight, in the early Saturday morning hours of February 8, a gigantic fireball raced across the Mexican sky and, accompanied by flashing lights, tremendous detonations, and a strong air blast, crashed to earth near the town of Pueblito de Allende, scattering a rain of meteoritic stones over hundreds of square miles. The local newspaper, *El Correo de Parral,* published an excited account which was picked up by the American wire services, and Elbert King of the NASA Manned Spacecraft Center in Houston heard about it Monday morning. By Wednesday he had been to Mexico and was back again in Houston, beginning work on the most important meteorite ever to fall on earth.

In Washington, D.C., Roy Clarke and Brian Mason of the Smithsonian National Museum were just a bit slower. They arrived in Mexico as King was returning to Houston, and spent the better part of a week there, investigating the fall and collecting thousands of grams of the most important fall since Satan's, bringing the pieces back to Washington for redistribution to laboratories all over the world. They were greeted rapturously; it seemed almost too good to be true. The meteorite, named after its place of fall, Allende, was one of the most rare types of meteorite, a C3 carbonaceous chondrite, and it was monstrous big. Most carbonaceous chondrites are pitiful little things which are so fragile that they break up in the atmosphere and fall to earth as minuscule black pebbles; searchers usually find a few little fragments which they wrap in their handkerchiefs and carry home carefully. When research scientists want to get a piece of one, there is much soul-searching before a few milligrams are handed over.

But Allende weighed two tons: there was plenty for everybody. All the meteorite workers who were gearing up to handle the lunar rocks suddenly had this rare C3 to practice on—and practice they did. Samples were generously distributed by Clarke to laboratories all over the continent and, indeed, all over the world. In laboratories in Fayetteville, Arkansas, and Rolla, Missouri, in Berne and Mainz, in Pasadena and Chicago, in Sheffield and Paris and Moscow and Minneapolis and New York and in both Cambridges, in more than thirty-seven laboratories in thirteen countries the lunar scientists tested their equipment with Allende—and found results more important than they would ever get with the lunar samples. The gods are devilishly clever in ways to drive men mad.

In 1965 H. C. Lord III had predicted what elements and minerals should be the first material to condense from a hot solar nebula as it cooled. Now the meteoriticists found in Allende small, whitish inclusions. Initial analyses showed them to be high in calcium, aluminum, and other refractory elements, i.e., elements that are condensable only at high temperatures; they were also quite low in volatile elements, i.e., elements that are not condensable at such high temperatures. Later that same year, 1969, the first lunar samples came back to earth, and were found to be remarkably similar: rich in refractories and depleted in volatiles. This suggested a remarkable and unlooked-for possibility: that we had serendipitously found in the moon and in the Allende calcium-aluminum-rich inclusions (CAIs) the first solids to form out of the original nebula. The lunar rocks were small and few and had cost billions of dollars to bring back, but the Allende meteorite was large and cheap—the Mexican farmers had been paid the ludicrous/immense (take your pick) amount of several thousands of dollars for their "windfall"—and researchers flocked to the Smithsonian to ask for samples.

They got them. There was enough for everyone, and the lunar resemblance stimulated everyone. By 1972 Larry Grossman of the University of Chicago was able to calculate the high-temperature condensation sequence in sufficient detail to show that the minerals found in the inclusions were precisely what the first compounds to condense out of the hypothetical hot solar nebula should be, as shown in table 21.1, which lists some of the refractories calculated to condense from a nebula at a pressure of 0.001 atmosphere within a temperature range of 1475° to 1450° K.

Table 2.1

Compound	Calculated Composition (%)	Measured Composition (%)
CaO	32–27	22–27
Al_2O_3	29–35	29–32
TiO_2	1.5–1.8	1.0–1.5
MgO	9–17	11–12

We had here sudden and unexpected evidence of the most convincing kind that the protosolar nebula had in fact existed, had been heated to high temperatures, and as it cooled had shed chemical condensates in a predictable thermodynamic sequence. Later work by Grossman and Clark emphasized the experimental-theoretical collusion: the most refractory minerals (those that would form first from the hot, cooling nebula) were actually enclosed in envelopes of minerals predicted to form later, at slightly cooler temperatures.

Everything was beginning to fall in place, it was a time of great excitement, the best of times. What to look for next? The answer was obvious: if these inclusions and the lunar samples are representative of the first condensates from the early solar nebula, they must be the oldest solid objects in the solar system. The chondrite meteorites in general are 4.6 billion years old—and almost before the sentence was finished the samples were in the mass spectrometers and the ages were clicking out of the computers—and the best of times turned once again into the worst of times, for the lunar samples were only 4.1 billion years old or less.

The moon, it turned out, was not the oldest object in the solar system, not the most primitive, not an early condensate direct from the solar nebula. The lunar rocks had not been formed by high-temperature condensation in the first moments of the solar system; instead they lost their volatiles in a later (unknown) event. Harold Urey was wrong, everyone was disappointed, the moon was going to turn out to be much more complicated than anyone had anticipated. Even today it resists telling a simple story, and (as we have said) we still don't know how it formed or where it came from.

But the CAIs in Allende, what about them? The first age estimates were encouraging, giving ages as high as any ever measured. The fact that they weren't higher, it was suspected, was due to our lack of sufficient precision in measurement: error estimates on chondritic ages were roughly plus-or-minus 0.1 billion years, and the question now was whether or not the first condensations in the solar system had come within this time interval. Amazingly enough, we knew that they had. That is to say, we already knew that the chondrites had begun to form within a hundred million years (0.1 billion years) of the end of whatever processes had created the elements of which they and we are made. The story of this discovery is certainly worth a chapter of its own.

SOURCES: Grossman 1972; Ringwood 1979

CHAPTER TWENTY-TWO

•

XENOLOGY:
THE AGE OF THE ELEMENTS

I N 1959 John Reynolds hit gold in his mass spectrometry laboratory in Berkeley, California. A professor of physics, he specialized in designing better mass spectrometry instruments for the measurement in geological materials of the rare gases, or noble gases.

Until then these gases had not been particularly important in scientific history. They were given the name "noble" in a rare instance of scientific sarcasm: like the nobility, they do no work. (The chemical elements enter into reactions via their atomic electrons, combining these in ways designed to reach satisfactory quantum mechanical levels; but the electronic orbits of the rare gases are already satisfactorily completed in their original state and so they are unable to enter into chemical combinations, and therefore unable to react to produce energy or to do anything else: another name for them is "inert.") They are also called "rare" because they are found in such small abundances on earth, although in the universe as a whole their abundances are perfectly normal. Helium, for example, the first rare gas, is more abundant in the sun than all the other elements combined, except hydrogen. This is shown in figure 22.1, where the abundances of the elements are plotted against their atomic weights.

This figure demonstrates one of the only two important geologic uses of the rare gases prior to Reynolds' work; it tells us something basic about the formation of the earth, in answer to the obvious question: Why are

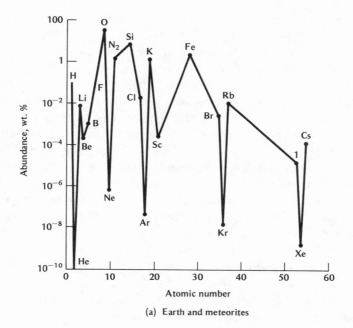

(a) Earth and meteorites

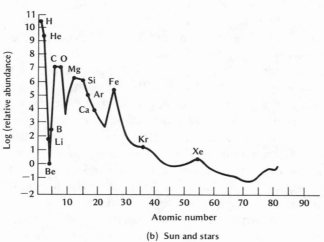

(b) Sun and stars

Figure 22.1.

the rare gases rare in the terrestrial planets (earth and meteorites) but of normal abundance in the sun and stars, when other gases, like nitrogen and oxygen, have normal abundances everywhere?

The answer must lie in the fact that the rare gases are gases under nearly all conditions, since they don't enter into chemical combination with other atoms. Nitrogen and oxygen can react with other elements to form solid compounds, nitrates, and oxides, but the rare gases are doomed (like the poor *Fliegende Holländer*) to a solely nebulous existence. Their low abundances in the terrestrial planets must indicate, then, that when these bodies formed they were unable to retain a gaseous atmosphere. Further, we know that their temperatures must have been low enough for normally gaseous species like oxygen and nitrogen to be locked up in solid compounds within the growing earth and meteorites. Sometime later in earth history, when our planet had grown massive enough to retain its gases, the temperature rose sufficiently to destroy these low-temperature minerals and release their gases, which bubbled up to the surface and formed our atmosphere. In the meteorites, particularly in the carbonaceous meteorites, these low-temperatures minerals still exist, indicating this high-temperature stage was never reached.

The second use of the rare gases was and is in radioactive dating. Isotopes of both helium (He-4) and argon (Ar-40) are products of the radioactive decay of species with half-lives on the order of the age of the earth: uranium-238 produces helium with a half-life of 4.5 billion years, uranium-235 produces it with a half-life of 0.71 billion years, and thorium-232 produces it with a half-life of 14 billion years, while potassium-40 produces argon with a half-life of 1.3 billion years. Since both these gases are so rare on earth, to a good approximation they can be considered to have been formed in particular rocks only by radioactive decay *in situ;* and sinc ? they are produced over a period of billions of years their measured abundances are good measures of the age of their host rock. We have seen earlier how Rutherford used the U/He pair to get the first good estimates of the age of the earth. During World War II a German scientist, Friedrich Paneth, who had come to England in the infamous year 1933 and who twenty years later returned to be a director of the *Max Planck Institut,* tried the same technique to determine if iron meteorites had formed as part of the solar system or if they were intruders from deep space.

His technique was based on measuring the uranium-helium ages of the meteorites (he had tried to measure argon, but his apparatus wasn't sensitive enough to disentangle the meteoritic argon from atmospheric contamination; argon constitutes 1 percent of our atmosphere, while helium is vanishingly small). His argument was that if the meteorites showed the same age as the earth, it would be reasonable to assume that they formed together as part of the same process; if they were older it would indicate that they were not part of solar system formation and must have come from elsewhere, while if they were younger one could say nothing about their origin since there is always the possibility that they might have lost helium by diffusion.

So it was a good experiment, since two of the three possible results would be meaningful in terms of solar system history. The age of the earth was then thought to be about 3 billion years, and Paneth's meteoritic data turned out to give uranium-helium ages ranging up to seven billion years. Which bothered people.

Something was wrong: it didn't jive, as people said in those days. If the meteorites were that much older than the earth, they must have come from beyond the solar system. But beyond the solar system is *nothing:* for millions of billions of miles there is just nothing at all, so where did they come from? Any other possible solar system is so far away that the likelihood of so many meteorites happening to reach us from there is infinitesimal; one or two might be possible, but there are thousands of meteorites. And the thought of iron meteorites forming individually and in profusion throughout empty interstellar space was nonsensical.

Or at least worrisome. But the year was 1942, and there were other things to worry about. Paneth became one of the leaders of the joint British-Canadian atomic research project, and it wasn't until after the war that he returned to the problem. While he was getting his laboratory at Durham University back to a peacetime footing, Carl Bauer in the United States and H. E. Huntley in England independently came up with the answer: the uranium-helium age depends on the assumption that all the helium comes from the radioactive decay of uranium. And that turns out not to be necessarily true: high energy cosmic rays, which by then were known to pervade space, could induce nuclear reactions in the meteorites, and one particular result might well be the formation of helium-4 as the result of the breakup of iron nuclei from the meteoritic mass. This complication doesn't exist on earth, where the atmosphere shields terrestrial rocks from cosmic rays, but the higher energy rays can easily penetrate several meters of iron, and most meteorites are smaller than that.

Seizing on this suggestion, S. Fred Singer, now a professor in Virginia but then a new physics Ph.D. attached to the Naval Research Office at the U.S. Embassy in London, and K. J. LeCouteur of the University of Liverpool, independently calculated that indeed the cosmic ray effect could account for a majority of the helium found in the iron meteorites, and predicted a new effect that would serve as a test for the effect: helium-3, a rare isotope, would be produced at about 30 percent of the helium-4 rate. Paneth's group immediately measured the He-3/He-4 ratio in their meteorites and found it to range from 0.25 to 0.35, whereas the ratio found from the decay of uranium is about 0.00000001.

So Paneth's experiment told us nothing about the time of formation of the meteorites, but opened a new line of research. By making estimates of the rate of production of helium in nuclear reactions, the "cosmic-ray age" could be calculated: the length of time the meteorites have been exposed to the cosmic radiation. In the meantime, other techniques based on the radioactive decay of uranium and thorium to lead and of rubidium to strontium did successfully measure the true age of formation of the meteorites.

These results showed that the meteorites formed at the same time as the earth, 4.6 billion years ago. The cosmic-ray ages turned out to be a good deal younger, mostly in the range of tens to hundreds of millions of years, which is explained by realizing that cosmic rays can penetrate only a meter or two of meteorite; if the meteorites had been formed as large bodies, kilometers in size, they would have been shielded from cosmic ray effects until they were broken apart. This all fits with an origin in the asteroidal belt, where occasional collisions both break up kilometer-sized bodies and disturb the orbits of the resulting fragments so that they may cross earth's orbit and eventually fall as meteorites.

This was the situation in rare gas and meteoritic research in 1959, when John Reynolds found something different. At that time the few isotopic measurements that had been done on various elements had shown that the isotopic ratios were always the same in earth and meteorites, except for known effects like cosmic-ray interactions and radioactive decay. And this fit with our idea that the material of the solar system was homogeneous, both earth and meteorites forming from a well-mixed proto-solar nebula identical throughout in isotopic composition. Reynolds extended these isotopic measurements to xenon, the heaviest stable rare gas, for two reasons.

First, xenon has an unusually large number of stable isotopes, nine of them, ranging in mass from xenon-124 to xenon-136. This provides an opportunity to examine the isotopic identicality of the earth and meteorites in fine detail. Second, one of the xenon isotopes, Xe-129, has an extinct radiogenic precursor, introducing an interesting concept.

Iodine-129, the xenon precursor, does not exist today on the earth or in meteorites, but once it did. It is radioactive, decaying to xenon-129 with a half-life of about 16 million years, which means that 16 million years after it was somehow created, half of it will have changed spontaneously into xenon. And every sixteen million years half of what remains decays to xenon, so that after, say, 100 million years there will be only 1.3 percent of the original left. After one billion years there will be too little left to be measured, and after four and a half billion years we can reasonably say that there is "no" iodine-129 left at all.

Now imagine the following scenario. No matter by what detailed process the solar system was formed, the elements had been previously created in stars and were somehow accreted into it. We don't know (or didn't know, before Reynolds) when they were created. Perhaps it was billions of years before the solar system formed. In that case the radioactive I-129 would have totally decayed to Xe-129 prior to solar system formation and the Xe-129 would have mixed homogeneously with all the other xenon isotopes that had been independently created; in this case the solar system would be formed from a cloud of elements isotopically homogeneous in xenon—sun, earth, meteorites, all would have the same relative proportions of Xe-129 in their xenon, though they might each retain vastly different total quantities of xenon.

But another scenario is possible. If the solar system formed within a few million years of the creation of the elements, it would form with some I-129 still in existence. Some solid minerals, according to their chemistry, would incorporate greater or lesser amounts of iodine, and therefore of I-129. Subsequently the I-129 would decay to Xe-129, and then the minerals high in iodine would show an excess of Xe-129 relative to the other xenon isotopes. Iodine is a relatively unimportant trace element in the cosmos and its geochemistry was largely unstudied and unknown, but if any solid body in the solar system could be shown to have a Xe-129 excess, one could conclude that it formed within a few million years of the creation of the elements; precise model-dependent calculations could give a precise time. If no such Xe-129 anomaly were ever found, the conclusion would be that the solar system formed at a time greater than a few hundred million years after the formation of the elements, when all the I-129 had already decayed.

The original emphasis on finding a fossil daughter of an extinct nuclide, as Xe-129 is described, was Harold Urey's. In 1955 he had been trying to understand how the first generation of pre-planet planetesimals could have melted. Such melting was needed to provide for chemical differentiation in his original scheme of planet formation by homogeneous planetesimal accretion. This scheme has by now long been discarded, but the fruits of the search it inspired still taste sweet.

According to his estimates the planetesimals weren't big enough to melt like the terrestrial planets did, by retaining heat from the radioactive disintegration of potassium and uranium. (Being smaller, their surface-to-volume ratio is higher, and they would radiate heat from their surface too fast to allow the temperature to build up). He suggested that if the planetesimals formed quickly enough after the elements had been created in some previous stars, there would still be "short-lived" radioactivities present, i.e., radioactive elements with half-lives on the order of millions of years. The basic law of radioactive decay is

$$A = 0.693 \times N/T_{1/2}$$

where A is the rate of radioactive energy production, N is the number of radioactive atoms present, and $T_{1/2}$ is the half-life: so the shorter the half-life, the more rapid the energy production. Short-lived radioactive elements, with half-lives on the order of millions of years, would generate energy at a rate thousands of times greater than do potassium and uranium, with their billion-year half-lives.

Of course, the energy production also depends on the number of radioactive atoms present in the planetesimals, and there wouldn't be enough iodine around for the hoped-for I-129 to have any significant heating effect (see figure 22.1), but the isotope aluminum-26 had just been discovered in 1954. Aluminum is one of the most abundant elements in meteorites and the terrestrial planets, and therefore in their parent planetesimals, and the Al-26 half-life is 0.8 million years: long enough so that there might still be sufficient Al-26 around when the planetesimals formed (if they formed quickly enough after element creation) and short enough to provide lots of heat. But there were a combination of theoretical problems (it didn't seem that there would be sufficient production of Al-26 in stars) and experimental problems (its fossil daughter is an isotope of magnesium, which is also a very abundant element; isotopic anomalies are most easily seen in rare elements, where a fossil contribution would stand out more clearly). And so no one searched seriously for Al-26's fossil daughter.

The I/Xe pair, however, looked more promising. Xenon is the rarest of elements, so any extra production of a xenon isotope would stand out clearly against a low background abundance, and iodine-129 would be formed by one of the normal processes of element synthesis in stars, the r-process, as described by B^{2}FH.

When we discussed stellar nucleosynthesis earlier, we mentioned that the heavy elements are built up in the late stellar stages by neutron capture. This occurs because incomplete fusion events or glancing nuclear collisions strip neutrons from the complex nuclei being formed, and this produces a sea of free neutrons within the star. The neutron, of course has no electric charge and therefore no Coulomb repulsive barrier between itself and the complex positively charged nuclei: when it encounters a nucleus it slips right in. As soon as it enters into the nucleus it is gripped by the intense nuclear force which is independent of electric charge, very strong and very short ranged (so that it doesn't operate until the neutron has actually slipped inside its borders). And a new nucleus is formed, for example:

$$^{16}O + n \rightarrow {}^{17}O$$

This process can continue again and again:

$$^{16}O + n \rightarrow {}^{17}O + n \rightarrow {}^{18}O + n \rightarrow {}^{19}O$$

but not without end. The sea of neutrons within a red giant star is a dilute sea, and on the average several thousand years will intervene between successive neutron captures. In the above example an oxygen-16 nucleus might capture a neutron, but then a thousand or so years will go by before it encounters another neutron on its wanderings through the star. And the resulting oxygen-17 will again live a couple of thousand years before it becomes oxygen-18 and again before it becomes oxygen-19. And before the process takes another step something else intervenes.

Stable nuclei are composed of protons and neutrons in roughly equal numbers, due to an exotic preference of the nuclear force. The nucleus can use — and, in fact, prefers — a slightly greater number of neutrons, and the bigger the nucleus the greater the neutron excess permitted, but there are limits. If a nucleus should have too many neutrons it becomes unstable: one of the neutrons spontaneously emits an electron and transmits itself into a proton:

$$n \rightarrow p^+ + e^-$$

The electron, which is sent flying out of the nucleus, did not exist before the process began; it is created at the moment of its expulsion. The process is called beta radioactivity (the electron is called a beta particle) and the resulting nucleus has one less neutron and one more proton. Since the number of protons defines the chemical elements, the new nucleus is a new element. In the above case the oxygen nuclei, each of which have eight protons and successively eight, nine, ten, and eleven neutrons, finally reach, with the creation of oxygen-19, an unstable, radioactive nucleus: a neutron transforms into a proton, the resulting nucleus has now ten neutrons and nine protons, and is an atom of fluorine (F-19):

$$^{19}O \rightarrow {}^{19}F + e^-$$

The half-life of the radioactive O-19 nucleus is only a few seconds, so clearly its radioactive decay will take place before it could possibly capture another neutron to become O-20. The nuclei can be plotted in a diagram of proton number versus neutron number, to see the effect of this alternation of neutron capture and beta decay in building up a succession of new elements (figure 22.2).

The process has its limitations, which are the result of the long time delay between successive neutron captures. In each case in figure 22.2 the radioactive half-life that interrupts the neutron capture process is short compared to the thousand-year interval between captures. If a radioactive

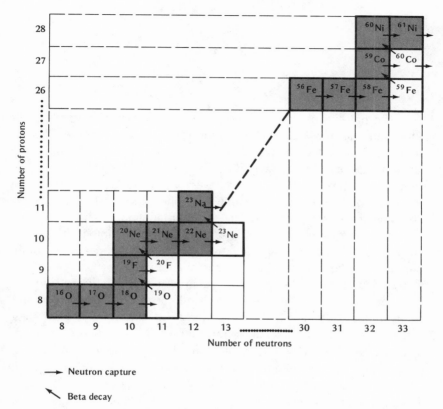

→ Neutron capture

↖ Beta decay

Figure 22.2. The shaded areas represent stable nuclides

nucleus is formed with a half-life long compared to a thousand years, it behaves in this process as if it were stable:

$$^{39}K + n \rightarrow {}^{40}K + n \rightarrow {}^{41}K$$

Radioactive K-40 has a half-life of 1.3 billion years, and so it is likely to capture a neutron long before it decays; the potassium chain continues on to K-42, which decays to calcium with a twelve-hour half-life, ending the production of potassium isotopes.

But consider the production of iodine:

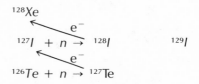

The half-life of I-128 is twenty-five minutes, a bit shorter than the thousand years it will take the newly formed I-128 to find another neutron, and so I-128 will decay to Xe-128—and the chain of nuclide production is broken without I-129 ever having been reached. So if this is the only method of nuclide production, I-129 would never exist. This is a problem for many heavy isotopes, including all uranium isotopes, and yet they do exist; so we know there must be another means of production.

A supernova is a cataclysmic explosion, dwarfing into insignificance the biggest H-bomb on earth and, indeed, the total energy output of the sun itself. In the supernova atomic nuclei are ripped apart, fragments are flung about, new nuclei are formed and destroyed, total chaos prevails, and neutrons come flying out of shattered nuclei in wild profusion, infinitely more abundant than in the sparse red giant sea. One of these neutrons slams into a nucleus to form a more neutron-rich isotope and within a microsecond another neutron may hit it, and another and another and yet another—all within the space of millionths of a second. Whether these new nuclei are too neutron-heavy to be stable, whether they are radioactive or not, doesn't matter—they haven't time to decay before another neutron slips into them and carries them further along the process. Finally their radioactivities catch up with the process, and what we have then is a group of exceedingly neutron-rich nuclei beta-decaying one after the other (figure 22.3).

The red giant neutron capture process is named the "s" process, for slow-capture; the supernova process is the "r" process, for rapid-capture. Between the two of them, all the isotopes of all the heavy elements are formed. Iodine-129 is formed only by r-capture, which takes place within the supernova explosion. And of course the explosion destroys the star and throws the atomic debris out into space virtually at the moment of its creation. If the meteorite parent bodies formed within a hundred million years or so of this supernova explosion, some I-129 would still exist; a mineral that trapped a reasonable amount of it and very little xenon would

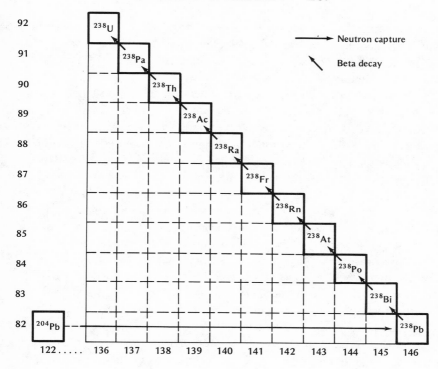

Figure 22.3. The r-process, starting from one typical nuclide (Pb-204) and ending with another (U-238).

today show an excess amount of Xe-129, compared to the other xenon isotopes, because of the subsequent decay of this trapped I-129.

The Xe-129 anomaly would be easier to find than the Mg-26 anomaly, primarily because of the low abundance of xenon and the longer half-life of I-129. In 1955 Wasserburg and Hayden undertook the search. They found nothing: the xenon in the meteorites they studied had the same isotopic composition as normal terrestrial xenon. The implication was that it had taken so long for the meteorites to form that any short-lived radioactivity had long since decayed away—which of course ruled out Urey's suggestion that such radioactivities were the heat source for his planetesimals.

And that, of course, made Urey unhappy. He suggested that palladium-107 (half-life = 7 million years) might be observed in iron meteorites, where palladium might be enriched relative to silver, its decay product. In 1957 he and his coworkers found nothing: no evidence of a silver-107 anomaly due to palladium decay.

These two unsuccessful experiments convinced most people that Urey's idea was wrong, that the time interval between creation of the elements and formation of the planetesimals was too long for the short-lived radioactivities to be present, and the searches petered out. Most rare gas experimentalists were happy to forget Urey's idea since the I/Xe experiment was so difficult. It was difficult because Xe is the mother of all rare gas difficulties: it easily adsorbs onto glass and metal surfaces and gets itself lost within the apparatus designed to measure it. Its abundance in terrestrial or meteoritic minerals is so low that it is hard to measure precisely—and very precise measurements were necessary because the 1955 null results implied that any isotopic differences that might exist would be quite small. The theorists, for their part, now found it hard to imagine the possibility of removing the elements from the stars, resolving them into a self-gravitating cloud, mixing them homogeneously, condensing them into a solar system (which would probably melt any solids previously formed) and then finally cooling them down to temperatures where permanent minerals would be formed, all within a time less than many millions of years. So not many people really expected a xenon anomaly ever to be found.

I was a postdoctoral in nuclear chemistry at Brookhaven in 1959. Working with the rare gases, I was concentrating on cosmic ray ages, extending the Paneth/Singer line of work. Together with Oliver Schaeffer I had just published one of the first papers on the determination of cosmic-ray reactions in iron meteorites when John Reynolds visited us and said that he had found a Xe-129 anomaly in the Richardson meteorite. I didn't believe him; the experimental difficulties were too great (I had spent a few months looking for meteoritic xenon variations and had found none) and the theoretical arguments against the anomaly were correspondingly convincing. (Theories increase their power to convince in direct proportion to the difficulty of the experiments necessary to disprove them.)

But Reynolds was right. His original data are shown in figure 22.4, with the horizontal line at mass 129 showing where the normal abundance peak would be.

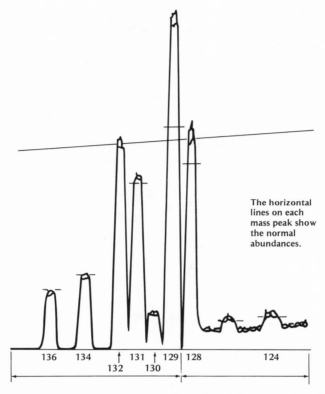

The horizontal
lines on each
mass peak show
the normal
abundances.

136   134   ↑   131   ↑   129│ 128   124
              132   130

Figure 22.4. Mass Spectrum of Xe from the Richardson Chondrite

Clearly there is an excess of xenon at mass 129. This one result transformed the field of meteoritics. A series of brilliant grad students and postdocs were attracted to Reynolds' Berkeley lab, and together they designed a number of experiments that established xenology as a scientific discipline in its own right. The measurements could in principle determine the time interval between creation of I-129 in a supernova and its incorporation in the first minerals to form in the protosolar nebula, but this would necessitate measuring both I-129 and Xe-129 today; given the present age of the solar system and the half-life of I-129, it is clear that no discernible primordial I-129 can be found today. Instead the experiments determined the amount of I-127, the stable isotope of iodine, and used theoretical arguments to estimate the I-129/I-127 ratio at the time of nucleosynthesis. The results are model dependent in their precise answers, but all agree (within a factor of two) that the meteoritic minerals formed within about fifty million years of the last nucleosynthetic event which produced extinct I-129.

This is a long period of time by human standards, but merely a wink of the cosmic eye. No longer was the formation of the solar system separated from the rest of the galactic processes by the impenetrable haze of uncountable billions of years. We were, for the first time since the ancient Hebrews conceived the idea of "In the Beginning," temporally linked to the rest of the universe.

SOURCES: Paneth et al. 1952; Reynolds 1960

CHAPTER TWENTY-THREE
•
THE CUCKOO'S NEST

REYNOLDS' DISCOVERY of the Xe-129 anomaly opened the flood-gates, and a vast trickle dribbled through. In 1960 Rama Murthy in California reported another isotopic anomaly, finding the effect due to the extinct decay of palladium-107 to silver-107 that had been missed by Urey a few years earlier, but by 1966 this analysis was shown to be wrong. In 1970 Brian Clarke and coworkers in Canada found a magnesium-26 anomaly due to the decay of aluminum-26; this also was later found to be wrong. For a decade after Reynolds' work the only significant isotopic studies were on the I/Xe system, showing that all chondrites had a similar time interval of about 50 million years between element synthesis and the formation of their minerals.

Time was not being wasted, however. During the 1960s a group was being organized at CalTech. Led by Gerry Wasserburg, one of the former Urey kids at Chicago, and calling themselves the Lunatic Asylum in antic-ipation of the forthcoming lunar samples, they were developing a new generation of mass spectrometers capable of both high sensitivity and ultrafine precision, and were matching these with newly worked out ultraclean chemical techniques. They were able to measure the magne-sium isotopes to an accuracy of 0.2 percent, and their first measurements on a variety of meteorites showed no anomalies at all; the solar system seemed to be isotopically homogeneous, in line with then current views. This first work, however, was a mere sharpening of their teeth, and when the Allende meteorite became available they pounced.

Their first great success depended on the radioactive decay of rubidium-87 to strontium-87, which occurs with a half-life of 47 billion years. Measurements of both Rb-87 and Sr-87 can yield an age, but other complications intervene and the measurements could not be made precise enough to detect an age difference of a few million years, which the I/Xe values had shown would be necessary to differentiate events at the beginning of the solar system. But the simple ratio Sr-87/Sr-86 could be measured with precisions of 0.2 percent at the Lunatic Asylum; the significance of the measurement is that this ratio will obviously increase in the early solar nebula as Rb-87 decays and Sr-86 remains constant, so that the earliest stuff to condense from the nebula will have the lowest initial Sr-87/Sr-86 ratio. The data measured today has to be corrected for the decay of Rb-87 *in situ* since that time of first condensation, but if this could be done the measured 87/86 ratio could be used to differentiate early solar nebula condensates from minerals formed later. Some of the Allende materials have a naturally high Sr/Rb ratio which makes this correction minimal, and with the improved technique that the Lunatics worked out they obtained results, suitably corrected, which gave for the Allende minerals the lowest initial 87/86 ratio ever measured, in exact accordance with the predicted scenario. This established that the material was the earliest known condensate from the solar nebula.

Meanwhile, back at the University of Chicago, another series of experiments was giving a totally unexpected and startling result. This part of the story actually began a few years previously, in 1969, at the University of Minnesota, but no one realized it. It began with a small experiment, little more than an anecdote about neon, one of the lighter rare gases. The isotopic ratio Ne-20/Ne-22 is 12.5, but certain carbonaceous chondrites showed ratios as low as 8.2. It was barely possible that such a variation could have been achieved by mass fractionation during gas loss: as a gas is distilled off from a solid, the lighter isotopes will leave preferentially, so the residue will have a reduced light/heavy isotopic ratio. One of John Reynolds' former students, Bob Pepin, currently at Minnesota, and his student, David Black, began a systematic study of neon in carbonaceous chondrites to pin down this possibility. They found instead that when they heated the meteorites through a range of temperatures they were able to separate out a component with a Ne-20/Ne-22 ratio as low as 3.4—and this was much too low to have been generated by mass fractionation (it would have led to an impossibly high initial content of neon in the meteorites).

It wasn't until three years later, in 1972, that David Black took the temerarious step of publishing a suggestion that this neon component (named Neon-E for historical reasons which are no longer of interest) was of *presolar* origin: that it is the result of a particular nucleosynthetic step that forms Ne-22 preferentially (or sodium-22, which decays to Ne-22), that it was incorporated in dust grains outside of the solar system (perhaps in the expanding debris of a supernova explosion), and that these dust grains were then blown into the interstellar cloud that was to form the solar system, maintaining their identity throughout the formation process so that their incorporated neon never was vaporized and allowed to mix with the rest of the solar system's gases. The suggestion is that somewhere in the meteorites measured by Pepin and Black there were solid grains that predated the earliest solar system, that had been incorporated in the meteorites without being melted or destroyed.

These hypothetical grains could not be CAIs which are formed of high-temperature refractories: the rare gases would not be trapped by any material condensing at such high temperatures. But if some low-temperature mineral (with Neon-E) survived the formation of the solar system, it was certainly possible that the CAIs did, too—except that no one who was anyone believed that the Neon-E was truly presolar. Exotic explanations for single data sets are not readily acceptable by the community. If presolar grains actually existed—if conditions in the protosolar nebula were not hot enough to vaporize any preexisting grains and mix them in with the rest of the gas before the whole lot began to condense—somehow and somewhere there would have to be more evidence than a single temperature-release analysis of a single rare gas. Exotic explanations have a way of coming and going without leaving their footprints on the sands of time, and people focused their attention on other matters while they waited patiently for this particular footprint to fade away. John Reynolds (Black's scientific grandfather) told him that if his (Black's) explanation was correct, isotopic variations should have been found in more common elements, such as oxygen. And no one had ever seen such exotic variations in the oxygen isotopes.

There was a slight flurry of interest in 1975 when K.K. Levsky and

A.N. Komarov in Russia found U/He ages ranging up to six billion years for iron sulfide grains in iron meteorites. They suggested that these grains were presolar; but their data were not strong enough to convince anyone, and I was recently able to show that they were the result of an experimental artifact in which the uranium and helium were separated, negating any age estimate made from them.

And then at the University of Chicago Bob Clayton and his group completed a long series of oxygen isotopic measurements in various types of meteorites, earth, and moon. Oxygen isotopic chemistry has long been a rewarding topic on terrestrial materials because oxygen is the most abundant element in the terrestrial planets, it forms volatile as well as refractory compounds, and it has three stable isotopes: oxygen-16, -17, and -18. Its high abundance means that it's easy to measure precisely; an experiment that measures the rare gas isotopes with a precision of a percent or two is very good, but the oxygen isotopes can be measured ten times as accurately. Chemical reactions that flux an element through volatile and solid states give the greatest opportunity for isotopic fractionation, since the lightest isotope will vaporize most easily and the heaviest will condense most easily. And finally, the existence of three stable isotopes enables one to make a meaningful analysis, to pick out simple chemical fractionation processes from anything more unusual or exotic, since for small fractionation effects (on the order of a few percent, as is commonly observed in geological samples) the dependence on mass of chemical processing is virtually linear.

The isotopic abundances are expressed in terms of measurements relative to a standard substance, as:

$$\delta - 18 = 1000(R^{18}/R^{18}_{STD} - 1)$$

where $R^{18} = {}^{18}O/{}^{16}O$, the subscript *std* refers to measurements made on the standard, and the data are reported in units of *per mil* (‰). The ${}^{17}O/{}^{16}O$ ratio is similarly expressed as $\delta - 17$, and then the two can be plotted against each other in a three isotope graph as shown in figure 23.1 for terrestrial, meteoritic (but not carbonaceous chondritic), and lunar samples:

In such a graph, if all the samples have the same nuclear history but different isotopic abundances due to simple chemical/geologic processes

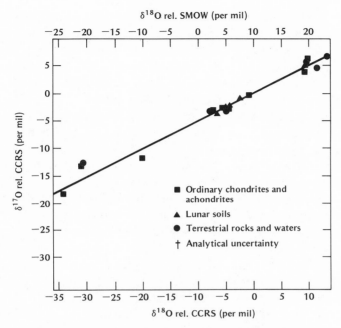

Figure 23.1. Oxygen in Meteorites, Earth, and Moon

having fractionated them, all the data points should lie along a straight line with slope of ½ (because the $^{17}O - {}^{16}O$ mass difference is ½ the $^{18}O - {}^{16}O$ difference); and indeed they do. The conclusion from this set of data is that the meteorites, the earth, and the moon have all evolved from a single primordial reservoir (the solar nebula) with one distinct set of original oxygen isotopic ratios, and that chemical and physical processes operating during their subsequent geologic existence have altered the ratios along the observed straight line.

Actually, the first attempt to look at oxygen isotope systematics in meteorites was way back in 1934, when Urey and coworkers searched for an indication that some meteorites might have come from beyond the solar system. They argued that such an origin might be reflected in an unusual relative abundance of the isotopes, as indeed it would be since the isotopic abundances are determined by the nuclear fires out of which the elements were created, and would be expected to be different in different solar systems which originated at different times and places in the universe. The precision of their analyses, in which they measured the oxygen – 16/oxygen – 18 ratio, was good to about 2.5 percent, and within this uncertainty they found no deviations from terrestrial materials.

With improved instrumental resolution in the 1950s and 60s, isotopic differences of a few tenth's of a percent were found, but these didn't seem to be terribly meaningful. In 1972 the Clayton group at Chicago reported that some minerals in Allende were up to 1 percent low in oxygen – 18, and suggested that this might be due to chemical fractionation between gas and dust in the early solar nebula. Such a model would predict a minimum δ – 18 of about − 12 per mil (1.2 percent), but later that year the same group found values as low as − 22 per mil in other carbonaceous chondritic minerals. They were momentarily bewildered. Looking back on it now, they might have wondered if they weren't seeing the same thing in oxygen as David Black thought he was seeing in neon; but they didn't.

In 1973 the group began a comprehensive series of analyses on the Allende inclusions. Larry Grossman had developed a model of solar nebula condensation processes that predicted oxygen – 16/oxygen – 18 variations due to thermal differences across the primitive solar nebula during formation of the minerals by condensation, and the object of the experiment was to investigate these predictions.

"For amusement", as Bob Clayton remembers it, he began to measure the δ-17 values which should of course correlate linearly with the δ-18 data. They did but, astonishingly, the correlation line they gave was different from the one previously observed. As Bob Clayton recalls:

> Tosh [Toshiko Mayeda] and Larry [Grossman] had bets on each sample as to the expected δ-value: Larry was looking for − 3‰, as predicted by the "cosmothermometer" for condensates at 1500K; Tosh was looking for correlation with the mineralogy, which she determined each day by X-ray. We were trying to concoct [invent] multistage fractionation processes to account for [some] δ-values [as low as] − 10 to − 20‰. Results came out high and low, until finally, on Saturday, May 19, we analyzed Al-16 and got − 26.2!
>
> I was alone in the lab, and called Tosh and Larry on the phone with the news. Because the sample had such low δ-18, I [looked for] the δ-17 effect. . . . On May 23 I plotted a graph of (δ-18 versus δ-17) and got a straight line, as expected, but the slope turned out to be [1 instead of ½]! I went home that night hoping that I had calculated something incorrectly, and that it would be all right in the morning. The next day I could find nothing wrong

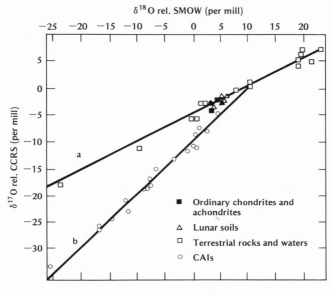

Figure 23.2. Oxygen in CAI

with my calculations, and began worrying about other people's. I checked
[the arithmetic and measurements of previous workers who had derived the
slope of the predicted correlation line and had done the first experiments
on terrestrial samples], but found nothing wrong.

For the next week they measured one CAI after another, including
minerals from other carbonaceous chondrites such as Grosnaja, Murchi-
son, and Vigarano, and by the end of the month they were convinced that
they had found something entirely new. Their data clearly defined a straight
line with slope of 1 instead of the mandatory ½ (figure 23.2).

Such a slope cannot be generated by simple chemical-physical processes dependent on the mass differences in the oxygen isotopes, and so
a different explanation must be found. There are two ways, in fact, to
generate a straight line in a three-isotope plot. One is the mass fractionation
effect we have discussed (which must give a slope of ½ for the oxygen
system); the other is to mix two components of different isotopic ratios
(which can give any arbitrary slope). In this graph, for example, if one
were to have instead of a suite of samples with continuously varying oxygen

isotopic ratios (due to mass fractionation) only two components, one with an isotopic composition which plotted at the point 0,0 and another which plotted at the point $-30, -30$, and if one were to mix these two components together in varying relative amounts, measurements of the resulting mixture would generate a straight line connecting the two end members— in this case, the CAI line.

The Clayton group suggested that this is what has happened. For one end-member they take the point 0,0 where the CAI line intersects the terrestrial-lunar-meteorites line: this represents a reservoir of oxygen identical to that in the "normal" samples. For the other end-member they suggest pure 0-16, which might be generated, for example, by helium-burning in the core of a red giant star or in the expanding envelope of a supernova. A mixture of these two compositions would produce the CAI line in figure 23.2. Presumably, then, interstellar dust grains formed somewhere out in space, incorporating pure 0-16 into their silicate structure (some form of $Ca-Al-SiO_4$). Eventually these dust grains mixed with the gas and dust that was to form our solar system, but due to their refractory nature they were never volatilized, never destroyed during the high-temperature processes that the nebula went through.

Back at the Lunatic Asylum, Wasserburg and his coworkers took another look at their magnesium isotopic results. Although they had found no anomalies in the wide group of meteorites that they had studied in 1970, the situation was now different: they had, in the Allende CAIs, material that they knew to be both ancient (from their own Sr-87/Sr-86 results) and to contain isotopic anomalies (from the oxygen data). With an increased sense of excitement they went back to looking for evidence of Al-26 in these primitive materials.

It took nearly another four years, but in 1973/74 the lunatics and an English team independently found verifiable Mg-26 variations amounting to roughly 0.3 percent in the Allende CAIs. This first evidence was difficult to interpret unambiguously because the effect was so small; it might have been due to a different (unknown) process depleting the CAI in the other magnesium isotopes instead of enriching it in Mg-26, but further work by two of Wasserburg's graduate students established larger effects and correlated them positively with aluminum content (i.e., the largest Mg-26 excesses occur in minerals with the largest aluminum/magnesium ratios) so the identification as the fossil daughter of Al-26 is complete.

Since then a host of isotopic anomalies in a whole range of elements have been found in the CAIs, damning forever to oblivion the old notion of a well-mixed, homogeneous solar nebula. But to make sense of the new, let's first remember the old. The idea was that the universe began with a Bang, expanding as a cloud of hydrogen and helium. Local condensations formed stars, and as these stars progressed through their lives they synthesized heavier elements in their interiors; when they exploded as supernovae they dispersed these newly synthesized elements throughout the interstellar medium. There they swirled around for immeasurable lengths of time, mixing continually with new injections of stellar materials to form a homogeneous medium. New generations of stars then formed from this mixture, the sun being one of these. As the sun formed it spun out a disk which clumped and aggregated into planets, chemically processing the initially homogeneous material but retaining the isotopic sameness.

But now we know that isotopic differences existed in the initial solar condensation. Furthermore, we know that the process took place when aluminum-26 still existed, which means within a very few million years of the formation of the aluminum-26, which occurred in the carbon shell of a supernova.

When massive stars, roughly ten times the mass of the sun, have exhausted all their nuclear fuel, they have a stratified structure. In addition to unused hydrogen and helium, they have concentrated iron group nuclei at the center, covered with successive layers of silicon-rich, neon-rich, oxygen-rich, and carbon-rich shells. Surrounding all this is a final primordial layer of helium and hydrogen. At this point, with no more nuclear fusion reactions pouring out energy to support the outer layers against gravitation, the star collapses. More accurately, the core collapses while the outer shells are blown off in a tremendous explosion during which their temperatures rise to billions of degrees. The energy is such that some of the nuclei are literally blown apart, giving rise to an intense flux of protons as well as neutrons.

In the exploding carbon-rich shell there is a quantity of magnesium-24 that had been created by the fusion of carbon ($^{12}C + {}^{12}C \rightarrow {}^{24}Mg$). This now can be hit by both a proton and a neutron to form aluminum-26. So as the supernova fragments are flung out into space, there is distributed throughout them freshly synthesized nuclei of Al-26. Four and a half billion years later we see the fossil evidence of this Al-26 as Mg-26 anomalies in the CAIs. It is interesting that the pure oxygen-16 required to satisfy the Chicago group's O-17/O-16 vs. O-18/O-16 correlations (figure 23.2) can also be produced in a supernova burst. In this case it is not so much a question of forming the O-16 in the supernova as of destroying the other two isotopes. Oxygen-16 is formed by one of the standard nuclear pathways, as discussed earlier, when an alpha particle fuses with carbon-12: $^{12}C + {}^4He \rightarrow {}^{16}O$. The heavier oxygen isotopes are formed by slow neutron capture (the "s" process) in the red giant stage or as side reactions accompanying hydrogen fusion, so all three of them are present when the star enters its supernova stage. The O-16, however, is a particularly stable nucleus while the other two are not, so that the cataclysmic conditions of the supernova preferentially destroy the mass 17 and 18 isotopes and can throw out nearly pure oxygen-16. (The O-17 and O-18 found in the solar system result from their expulsion from stars before the supernova stage, blown off from stellar surfaces by red giant solar winds.)

There is an interesting conundrum in this picture, relating to the age scale. The original isotopic anomaly, remember, was Xe-129, and it defined an interval between nucleosynthesis and meteorite formation of about 50 million years. The Mg-26 anomaly, on the other hand, must have been produced within 1 to 2 million years (the Al-26 half-life of just under a million years means that none would be left in 50 million years). The separate Mg and Xe analyses therefore give two distinct answers where one would do for the question: What is the time interval between creation of the elements and mineral formation? Fifty million years according to Xe-129, but 1 or 2 million years according to Mg-26.

Interesting puzzles demand interesting answers. One answer proposed for this puzzle is that any Al-26 produced together with the I-129 in the supernova that went off 50 million years before solar system formation decayed away long before it could be incorporated into solid minerals. Then, just 1 to 2 million years before the solar system formed, another supernova exploded, this time forming Al-26 but not I-129. In this way we find the fossil traces of I-129 from the earlier supernova and Al-26 from the later one.

This explanation seems a bit ad hoc, but it may be that supernovae come in different flavors, so to speak: some produce lots of r-process I-129 and others don't. We don't really know a helluva lot about supernovae, so who can tell? On the other hand, there is an interesting consequence of the idea:

Given that both Al-26 and O-16 are formed in the supernova, is it not a rather curious coincidence that the solar system began to form within such a short period of time after its occurrence? Although supernovae occur two or three times every hundred years within our galaxy, the galaxy is a huge place containing a hundred billion stars. The chance of a supernova going off at any particular point in space and time is remote. For example, we know from the geological record that no supernova has gone off close enough to us to influence us for at least the past 65 million years. (The Cretacious/Tertiary transition, mentioned in chapter 16, may just possibly have been caused by a supernova at that time, although the odds seem to be against it.)

The answer is that it is indeed a curious coincidence, unless the principle of cause and effect is operative: unless the supernova that created the Al-26 and O-16 also caused the solar system to form. This hypothesis was actually suggested several decades ago, before there was any isotopic evidence linking our solar system's birth to a supernova, when Fred Hoyle and Ernst Öpik independently took up the problem of how to get an interstellar cloud to begin its gravitational collapse into a star. The problem is that for such a cloud to exist, the mutual repulsive forces of its constituent particles (arising from their energy of random motion within the cloud) must balance the total gravitational energy; for it to collapse, the gravity must become dominant. One way for this to occur is for an outside force to induce a momentary compression of the cloud. If the atoms within the cloud are thus pushed closer together, their mutual gravity increases, pulling them even closer together, and the effect snowballs into gravitational collapse. Hoyle and Öpik suggested that the shock wave of an exploding star might be this outside force.

Putting it all together, we can envisage a cloud of gas and dust passing through a high density spiral arm of the galaxy 4.6 billion years ago. This impact compressed parts of the cloud into massive stars which raced through their life cycles in a few million years, at the end of which they exploded as supernovae. Remnants of the original cloud which did not collapse were still tumbling around, and one such cloud remnant was close enough to the blast wave of one of these supernovae to be squeezed into a density high enough to begin collapse into what became our solar system.

This concept of a supernova trigger to our formation is a lovely one, but it became less necessary a few months ago when data from the High Energy Astrophysical Observatory and the Solar Maximum Mission satellites showed the continuing presence of aluminum-26 in interstellar space, distributed throughout the galaxy. This implies that the isotope is formed not only in supernovae explosions but also in processes more common in galactic history, in ordinary novae or red giants for example, and is continually flushed out so that whenever a cloud collapses and begins the star formation process it will necessarily incorporate this radioactive isotope. It remains to be seen whether enough Al-26 is formed by such processes, or whether the meteoritic values do reflect the presence of a supernova. Either way, clearly the collapsing cloud did tumble and spin, speeding up as it contracted, to form a flattened sphere at the center of a surrounding disk or nebula. We must now ask how such a nebular disk could gather itself into a few distinct planets.

SOURCES: Wasserburg and Papanastassiou 1982

CHAPTER TWENTY-FOUR

•

AGGREGATION

THE PROCESS OF the aggregation of a large number of small planetesimals into a small number of large planets is particularly amenable to explication by computer simulation, since it is by its nature an
iterative accumulation. One starts with a swarm of bodies in reasonably
well-constrained orbits, the conditions of collision which will lead to
accumulation rather than to further fragmentation are established on
grounds of classical mechanics, and one then simply repeats the calculation under various assumed conditions to see if there exists a set of
reasonable conditions that result in a solar system similar to the one which
exists.

The first computer simulations revealed an unexpected problem. It
had been thought that planetesimal accretion could occur only if collisions
occurred with sufficiently small velocities; at higher velocities the colliding
bodies would mutually fragment and bounce apart rather than sticking
together to fuse into a larger body. For example, two boulder-sized rocks
colliding at six hundred miles per hour will break into pieces and fly apart,
but if the collision takes place at a much lower velocity they could in
principle coalesce (in the absence of other forces acting on them, such as
the gravity of the earth). The quantity important in relating the necessary
velocity limit to the masses of the colliding bodies is what has become
familiar to us today as the "escape velocity."

When we send rockets to the moon or to Mars, or to anywhere beyond
Earth, we must impart to them a velocity sufficient to allow them to escape
Earth's gravitational hold: the escape velocity, the minimum relative veloc-

ity between two bodies that will overcome their mutual gravitational attraction. For the rockets to escape from Earth this is roughly 11 kilometers per second, but when they lift off from the moon on their journey home they have to attain only 2.4 km/sec because the moon is a less massive body than Earth. Obviously the escape velocity is a function of mass: the smaller the bodies, the smaller the escape velocity.

This was the initial problem. The planetesimals are very small to begin with, millimeter to meter-sized, and so the escape velocity is also small. If two planetesimals collide with a relative velocity greater than that, they will fragment and fly apart; if less than that, though they may fragment, the fragments will be gravitationally bound and thus a larger body is built up. The extreme examples are two rocks colliding, smashing, and the pieces flying apart; or a meteorite falling into the earth, smashing and fragmenting, even volatilizing, but with all the meteoritic particles gravitationally bound to earth and thus falling back eventually to its surface. In this latter case the earth accumulates the debris; in fact we grow by many tons every day through this process. In the real world some of the collisional energy is dissipated through internal deformations of the colliding bodies and some is inevitably lost through heat, so even collisions at slightly greater than the theoretical escape velocity will result in aggregation, but this factor is not terribly large, a factor of two or three at most. The immediate problem, then, to begin planetesimal coalescence, is to get them to nudge each other gently at relative velocities below about 0.1 km/sec rather than bashing each other apart in more violent collisions.

This is done easily enough by putting them into similar orbits. Kepler's laws of planetary motion, which apply to any particle bound by a central force and therefore equally well to the final planets and to the initial planetesimals in orbit around the protosun, insist that a body's velocity is determined solely by its distance from the sun at any point, which is in turn determined by the shape and size of its elliptical orbit. So if two bodies are in similar orbits they will have much the same velocity vectors and therefore small velocities relative to each other; when they collide they will stick together. This is the solution to the first problem—and the essence of the second.

Consider a nebular disk in which all the planetesimals have concentric orbits; that is, orbits of closely similar shape, differing only in radial distance from the sun. Then those planetesimals of roughly the same distance will collide gently and grow, and other planetesimals of slightly greater or lesser radial distances will likewise collide with each other and grow—and then the process will stop, not because any further collisions would be too violent but because they will not occur at all. All the planetesimals of similar orbits will have collided, and all with dissimilar orbits never come close to each other. We end up with a large number of slightly greater planetesimals than we started with, traveling now in concentric, nonintersecting orbits, with no chance at all for further collisions and further growth. This is certainly not the planetary system we see today.

The number of bodies remaining when the orbits become nonintersecting, which terminates the aggregation process, depends on the allowed relative velocities. If the velocities are very small, aggregation will be most efficient; but the necessary condition for this is quite similar orbits, and the result is a large number of small bodies. Larger relative velocities allow more dissimilar nonconcentric orbits with consequently greater overlap and intersection possibilities, but the orbits cannot be too dissimilar for the velocities must remain below the escape velocity for the masses concerned. Obviously the best choice of a relative velocity is one close to the escape velocity, i.e., orbits as dissimilar as possible, providing the greatest overlap, which in turn means the highest velocities allowed rather than the lowest. And since the escape velocities increase as the planetesimals aggregate and their individual masses increase, in the best of all possible worlds one would wish for a scheme in which the relative velocities somehow keep step with the increasing masses.

In 1950 such a scheme was introduced by A. I. Lebedinskii and L. E. Gurevich, and developed in the 1960s by V. S. Safronov, all at the Institute of Applied Geophysics in Moscow. They recognized that while the initial velocities of particles in sun-bound orbit are determined by the shape and size of that orbit, two processes that occur during the planetesimals' continued revolutions further influence their motions. When two such planetesimals collide they either stick together or they do not. If they do not the collision tends to make the resulting orbits of the two bodies more circular than before, and therefore more similar: this effect will decrease the relative velocities of the planetesimal swarm. If two planetesimals pass close by without colliding, however, they nevertheless influence each other through their mutual gravitational attraction: the orbit of each of them is changed—they are slung about—and the result is that on the average their relative velocities toward the next planetesimal they will encounter is *increased*.

Since the slowing-down of velocities occurs only when planetesimals actually collide while the increase occurs when they pass close by, and since they are more likely to come close than to actually hit, the net result is that the orbits are continually modified in such a way as to increase the relative encounter velocities. In fact, as the individual masses of the planetesimals grow through the aggregation process, their effective gravities follow suit. This increases the velocity-increasing gravitational perturbational effect, with the lovely result that as the masses grow—and with them the escape velocities—the encounter velocities keep step.

At the beginning, then, the revolving and colliding bodies are small. Collisions between them tend to make the orbits circular and similar, reducing the relative velocities below the escape velocity and allowing them to begin to aggregate. As they grow their gravity increases, perturbing nearby planetesimals in similar orbits into dissimilar orbits, increasing their relative velocities but causing more intersections, and as the process continues the planetesimals grow and sweep up larger and larger areas of space. In this way we end up with a few large bodies instead of many small bodies: we end up with a solar system.

But do we end up with the solar system in which we happen to live? The aggregation process is by its nature a statistical one, and is not constrained to a single-valued solution even if we knew how to carry it out without simplifying assumptions. It does not necessarily lead to a system of precisely nine planets in stable, nonintersecting orbits at Titius-Bode distances. But before we try to answer the obvious question of just how probable a nine-planet T-B solution is under the postulated conditions of planetesimal aggregation, we must realize that we have come up against a tougher problem: we have no idea how probable our particular solar system is in the real universe, let alone in our imaginary calculations. It may be that most planetary systems have greater or fewer numbers of planets, with size, density, and distance distributions similar to or different from those in our own system; until we know, until we have seen several other systems and measured their characteristics, comparisons of theory with observation are of limited use. We must know just how typical our system is before we can with confidence go much further in theoretical origins and developments.

As an extreme example, if we should find that in fact we are alone in the universe, that other stars do not have their own planetary systems, then this whole concept of planet formation as a natural consequence of star formation would have to be deemed invalid. We should then divert our attention to more intrinsically improbable mechanisms for our own origin, such as Woolfson's stellar encounter hypothesis.

In the meantime the calculations continue to grow in sophistication. The original model was, despite its high levels of cleverness and difficulty, much too simple a description which did not do justice to the complexity of the aggregation process. Of course one cannot fault the Moscow theorists if their initial theory was incomplete: we are few of us Newtons or Einsteins, whose concepts seemingly step forth as Venuses born whole and perfect. Their initial simplifications were reasonable and necessary; they assumed first that the beginning planetesimal swarm consisted of equal masses, then modified that to the idea that the greatest portion of mass might be concentrated in the few largest bodies; the theory treated the planetesimals as if they were molecules of a perfect gas, in linear motion with no dependence on distance from the sun; and it was a two-dimensional model.

These simplifications have been more meaningfully developed by Richard Greenberg at the Planetary Science Institute and George Wetherill at the Carnegie Institution, with very encouraging results. It is a commonplace in science that one begins an assault on a new idea by making simplifying assumptions, but all too often it turns out that though the initial calculations are promising enough to generate years of effort introducing real-life complexities in order to haul the theory willy-nilly out of the imagination and into the real and physical universe, the theory then dies of such exposure like a vampire in the light of the everyday sun, incompatible after all with reality.

In this case, however, it seems that such an initial swarm of planetesimals as originally visualized, one that might have been formed by consolidation of dust grains in a protosolar nebular disk, might very well develop by mutual collisions into the sort of planetary system we see today. Indeed the situation is no longer one of trying to imagine how the planets conceivably might have formed, but of trying to devise experiments to decide between two reasonably viable alternatives—planetesimal accumulation into growing planets and gravitational gaseous collapse into huge protoplanets that later shed much of their mass. Although imagination and hard work, the two historical components of scientific progress, remain mutually necessary, the emphasis on them now appears ready to change from the former to the latter, both in the theoretical calculations and in the physical experiments which may soon bring us news of other planets circling distant stars.

SOURCE: Wetherill 1981

CHAPTER TWENTY-FIVE
•
THINGS INVISIBLE
TO SEE

S O THERE WE ARE.
    In a great primeval explosion some twenty billion years ago our universe began. Clouds of hydrogen and helium condensed out of the expanding debris and collapsed into stars, their gravitational energy heating them to millions of degrees until finally their constituent nuclei began to fuse together and liberate nuclear energy. They fused hydrogen into helium, then helium into carbon and then on into oxygen and neon and more; neutrons sailed through and among them and created one by one more complicated elements, different groupings of neutrons and protons, until one day this complex gallimaufry of atoms was flung out into space. There they drifted, dispersed, gathered into new clouds, and fell together again to form new generations of stars.

About fifteen billion years after it all started, just 4.6 billion years ago, one particular cloud began to collapse, spinning and tumbling. Its gravity pulled it into a spherical star, its angular momentum generated a disk around the star; as the angular momentum braked down the central star spun less slowly, and as the dust grains spun down into the disk and began to aggregate into planetesimals the solar system was formed.

There is hardly a single sentence in all this book that will not find its own particular antagonist who will rise up to deny it, to argue for an alternative thought, to point out a conflicting datum or calculation; but

taken together they all express an integrated concept of planetary formation that is a natural extension of universal stellar processes observed to
be operating yesterday and today. Taken together, they describe how the
earth and the planets formed.

Unless.

Unless those universal processes do not in fact extend as far as we
suppose. If we were disembodied intelligences existing without a home,
circling a star without planets, there would be nothing then in our scientific
studies to tell us that such things as planets must exist. Nothing in stellar
structures or evolutionary necessity stipulates that planet formation is a
necessary and unavoidable consequence of the existence of stars. But
rather we conceive of planets because we see them, and it is our triumph
to have constructed a broad theoretical model by which such planets
could be formed as part of the stellar sequence reasonably, naturally, and
without any leaps of the imagination. Which is not quite the same as
saying it *had* to happen this way.

In order to make the statement we need to expand our horizons, we
need to observe other planetary systems and see that they are indeed
natural outcomes of our universe, prevalent or at least not uncommon
companions to the multitude of stars; we need to compare their structures
to our own and to our models, to see their development in times both
great and small compared to our own; we need, first of all, to find these
things currently invisible to see.

As we wander through the galaxy at our own particular rate we pass
other stars moving slower and are in turn passed by stars moving faster.
From our apparently stationary planet we see these stars in relative motion,
appearing to move in straight lines relative to the starry background. But
if such a star is part of a multiple system, superimposed on this linear
motion through the galaxy is its orbital motion around its companion(s).
We see this clearly in binary star systems: each star appears to wobble
along its straight-line galactic path, oscillating as it circles its companion.
Binary systems are often discovered this way; an apparently single star is
seen to wobble, closer examination is made, and a dimmer star is then
seen as its companion. The same effect would be seen if the companion
were merely a planet rather than a star; the planet would be smaller and
therefore harder to detect, but it would be there.

In 1937 Peter van de Kamp began a decades-long experiment searching for such evidence of other planets. He focused on the closest star, Barnard's Star, less than six light years away, measuring its position as accurately as he could for year after year, searching for a wobble in its path which would be evidence of the gravitational tug of an unseen companion. Twenty-five years later he concluded that he had found it. Ten years later still, Robert Harrington at the United States Naval Observatory concluded that he had probably not found it.

The problem is one of precision. Current techniques are capable of measuring a star's position to an accuracy of about one arcsecond. (Divide a circle into 360°; each degree can then be divided into 60 arcminutes, and each arcminute into 60 arcseconds.) An observer on Barnard's Star, looking at our sun, would have to detect variations in its position of less than two-*millionths* of an arcsecond in order to see the effect of earth's gravitational pull. Jupiter's pull would create about two-thousandths of an arcsecond wobble. Here on earth our measurements of other stars' positions are good to about one arcsecond, nowhere near good enough to see an earth-sized or even a Jovian planet around Barnard's Star.

Van de Kamp's analysis attempted to, in effect, improve the precision by adding together a long series of analyses. The problem is that errors are also additive, often unaccountably, and so his conclusions have a slack in them that renders them less than totally convincing. (It doesn't help that the two largest effects he saw, on which his argument largely hinges, were observed in 1949 and 1957, which is also when his telescope was taken down for periodic adjustment. One can't help wondering if the coincidence is fortuitous or causal.)

In 1984 Donald McCarthy and Frank Low of the University of Arizona and Ronald Probst of the National Optical Astronomy Observatories announced the discovery of a planet circling van Biesbroeck 8, a relatively dim star twenty-one light years away. Their method was to look in the infrared region of the electromagnetic spectrum rather than in the visible; as we have mentioned, solid particles absorbing energy from a star reirradiate it primarily in the infrared, and so a dim planetary companion could be seen next to a star more easily in this wavelength region than in the visible, where the starlight would overwhelm it. They also thought to look at the closest *and* dimmest stars, to further minimize interference from the parent star. Using the technique of speckle interferometry, they found definite evidence of a planet circling about a billion kilometers around the star.

This time there is no doubt that the stellar companion is actually there. But is it a planet? It is several dozen times as big as Jupiter, and since van Biesbroeck 8 is much smaller than the sun its "planet" is not much smaller (less massive) than the star it circles. What is the difference between this pair, then, and an ordinary binary star system? It comes down to the question of the difference between a star and a planet, which is simply a difference in mass. As an interstellar cloud condenses to form a star, it heats up; the point at which the heat generated by this gravitational collapse is sufficient to turn on the H—He fusion furnace defines the birth of the star. If the object is not sufficiently massive, the temperature in its core simply doesn't get hot enough (a few million degrees) to ignite; the object may glow for a while, but will soon cool off into what theorists have been calling a brown dwarf, essentially a failed star.

Which is what van Biesbroeck 8's "planet" is. If it were just a bit bigger it would have sufficient mass to form a companion star. So it's interesting, it extends the observations to include a new class of previously only theoretical objects, but it has little to do with the question of the existence of extrasolar planetary systems. The question of whether it's a planet or a sort-of star is semantic, not scientific. What we really need to know is whether multiple small bodies can form around stars, which means detecting bodies at least as small (!) as Jupiter; much better would be to find Neptune-sized objects, and of course terrestrial-sized objects would settle the question definitively.

But this last observation isn't likely to be made soon. We don't quite have the techniques or equipment to do it. NASA is funding design studies for a series of experiments capable of detecting other-world planets, but none of these will be precise enough and sensitive enough to detect earth-sized objects. There are basically three techniques currently envisaged.

Two of them are based on the fact mentioned above: that all objects gravitationally bound revolve not around the central star, but around their common center of gravity, the central star also obeying this motion. Since in a solar-type system the central star includes most of the system's mass, its position coincides closely with the center of gravity, but not quite: it moves, affected by the tug of its planets' gravity. One type of experiment, similar to van de Kamp's Barnard Star planetary search, is designed to look for this motion. The problem is that precisions of 0.003, 0.0009, and 0.000002 arcseconds are necessary to detect planets of Jupiter, Neptune, and Earth masses respectively, at the distance of the closest star, and our present limits of precision are just about one arcsecond—and of course the necessary precision increases as we look at more distant planets.

An alternative scheme is to look at the Doppler shift as the star, circling its planet, moves alternately toward us and away from us. Precisions in wavelength measurement are much greater than in the fixing of stellar positions, but this effect will be seen only if we happen to lie close to the plane of the star's ecliptic, which will statistically eliminate most stellar systems. Current precisions are just about sufficient to pick out planets the size of Jupiter, and improvements to Neptunic masses are not beyond the current horizon, but it's not yet completely clear that nothing except planets could cause such motion: the interpretation would have to be intrinsically single-valued before the experiment would be worth much.

Currently the most promising techniques are the one used on van Biesbroeck 8 (infrared speckle interferometry) and the Large Deployable Reflector, an infrared satellite observer similar to IRAS but with higher resolution. The latter instrument could do the job, depending on its funding; the instrument envisaged with the current budget would be useful only for the very nearest stars. A positive result would settle things, of course, but a null result on just these few stars would not be meaningful. Other ground-based interferometric techniques are being worked on, and an orbital astrometric telescope is in the design stage. This latter instrument could do the job if the experimental problems are licked, which seems likely by the 1990s, and if funding becomes available, which seems likely about the time of the Second Coming (unless someone can think of a military application for it).

In the meantime there are two space systems scheduled for launch within the next decade: the European Hipparchos satellite and NASA's Space Telescope. These should be able to see, by a variety of techniques, Jupiter-sized planets around the very nearest stars; and that isn't much of an improvement over earth-based techniques and really isn't good enough. The telescope will probably do most of its best work in completely unrelated astronomical areas—there is already a long line of astronomers waiting to use it for their own projects.

Improvements in experimental techniques will undoubtedly continue to occur, but probably slowly; there is an ingenious and hard-working group of men and women striving toward planet detection, but the group is comparatively small and ill-funded. Part of the reason for the lack of urgency on the part of the rest of the scientific community is a feeling that we already know what the answer is: certainly there must be planets out there. We see so much of the process already that it is inconceivable to most scientists that it might stop with a sudden lurch short of forming actual planets around many stars. We see clouds of gas and dust condensing around new stars; we see infrared clumps of emissions in these collapsing dust clouds; we see binary stars and tertiary stars; and finally, in 1984, we saw a brown dwarf (or a large planet) around van Biesbroeck 8. Surely it doesn't stop there, surely the process continues in this most inefficient of all possible universes, surely around each forming star there is a band of leftover crud which may fall together or clump together or somehow come together to form planets which will someday be visible. Surely all this is true—but until we see another Jupiter or Neptune or even, some day, another Earth circling a foreign star far away, we cannot *know*.

And yet, for most of us, the certainty of those planets' existence is taken for granted; before we have yet seen their reflected light, we have lifted our eyes to further horizons, to the even more intriguing question: is there life out there on any of those planets, and—finally—is there intelligence out there, somewhere, elsewhere?

SOURCES: Black 1984; McBride 1984

CHAPTER TWENTY-SIX
•
GO AND FIND A FECUND STAR

THE MAIN IMPETUS of both theoretical and experimental work on the origin of our earth and solar system leads to the conclusion that they have formed as the result of natural processes accompanying star formation. If this turns out to be not true, if our earth is unique in the universe, then obviously we as intelligent living creatures are alone in this awesomely immense vehicle of nearly empty space-time. But if our ideas are correct, they raise the question of the existence of other forms of intelligent life elsewhere. Is the question a reasonable one?

From the beginning of time until 1828 there was a magic line that could not be crossed: it marked the boundary between living and nonliving things. The structure of all things, alive and sentient or inanimate and dumb, was obviously chemical—but the two chemistries were different. The chemical compounds found in living creatures ("organic" chemicals such as hemoglobin, insulin, chlorophyll, and urea) could be formed only in and by the living creatures themselves, although they were composed of the same eighty-two stable chemical elements that formed the rest of the material world (with the radioactive ones present in trace amounts). The proportions were different, 99 percent of biogenic compounds being composed of the four elements carbon, nitrogen, oxygen, and hydrogen instead of the iron, aluminum, magnesium, and silicon that (together with oxygen) are more common on earth, but that was hardly enough to explain the qualitative difference between living and dead things. Obviously there was something missing from the chemical analysis that could be provided only by a divine creator: the spark of life, the life force, the soul. This missing ingredient provided an absolute, insurmountable barrier between the two worlds: life was life and not was not, and never the twain could meet.

And then in 1828 Friedrich Wöhler in his laboratory at the University of Berlin mixed in a beaker the two simple inorganic compounds ammonium cyanate and water and heated them, boiled them, slowly cooled them, and analyzed them—and found $CO(NH_2)_2$—urea. And our world has never since been the same.

This was one of the most important scientific experiments of all time. It forever changed the vision we saw when we peered out from our skulls at the universe beyond, and yet so few people know about it. That one simple experiment burst irrevocably the thongs of magic that bound us, that blinded us, that kept us in thrall to the mythical kingdom of god. Urea, the basic chemical compound in our urine, became not only an ever-present reminder of the inefficiency of our bodies but a symbol of the power of our minds. For the first time we had synthesized, created, a biological compound; we had shown it to be identical to the compound created in our bodies; we had removed the element of magic from the concept of life. The field of organic chemistry was founded on that day, and sprang into a life of its own so quickly, so frenetically bringing forth discovery after discovery as one after another the chemicals of life were created and studied in the laboratory, that just a few years later Wöhler helplessly wailed: "Organic chemistry drives me mad! It's like a primeval tropical forest full of the most remarkable stuff—a dreadful endless jungle into which you'd better not enter because there seems to be no way out!"

Aside from wresting the art of medicine out of the hands of well-intentioned but hopelessly lost wanderers and bringing it to the edge of fruition as a science, this work of Wöhler's allowed the slow emergence of a rational concept of life itself. Slow it was, but emerging it is. More than a full century later, in 1953, Harold Urey and one of his graduate students, Stanley Miller, mixed together a beakerfull of the simple volatile compounds that might have formed the primitive atmosphere of the earth—water, ammonia, methane, and hydrogen—passed an electrical discharge through them, and induced chemical reactions that produced the most basic organic molecules—amino acids, the building blocks out of which all life is fashioned.

This is not to say that life was created in that Chicago laboratory. Life has never yet been created nor seen to occur spontaneously in nature, although at first glance it did seem to occur in this manner. It was Louis Pasteur in the nineteenth century who showed that the spontaneous appearance of maggots in dead flesh could be prevented if the dead animal was isolated from air, that the maggots were contamination from outside rather than a spontaneous manifestation from within. This was strong support for the magical creation of life. If life can not arise from nonliving things in the rich, beneficient world we have today, how could it possibly have done so on the primitive earth which must have had a much less abundant supply of oxygen, fertilizer, and possible nutrients (since all of these are the result of life itself, and so must have been scarce or nonexistent on the early earth)?

The answer to that lies in the question itself, for both oxygen and life are detrimental to living things. Organic compounds are easily oxidized, destroyed by chemical reaction with oxygen, liberating energy in the process for the benefit of any organism that knows how to control the reaction but simply destroying simple organic molecules that don't already have the knack of protecting themselves. The composition of the early atmosphere of the earth is still largely unknown, depending on such factors as the degree of equilibration with metallic iron during outgassing from the convulsing earth, which in turn depends to a large degree on the (unknown) rate of outgassing from the earth. One would like the early atmosphere to have been as reducing as possible, since such an atmosphere would have been conducive to the synthesis of organic compounds, but the possible range of atmospheric compositions is wide. Certainly water and hydrogen were important constituents, together with carbon in some form (CO_2, CO, or CH_4) and nitrogen (as N_2 or NH_3); just as certainly free oxygen was extremely scarce or even totally absent. Once the first organic molecules were formed (by nonlife-related chemical reactions such as the Miller-Urey synthesis) they could begin to accumulate, destroyed neither by oxidation nor by being metabolized by living creatures—the other fate of organics on today's earth. On that primitive earth, unlike today's, there was nothing to destroy the first organic compounds.

And so, born of lightning flashing through a simple atmosphere of water, hydrogen, and gaseous compounds of carbon and nitrogen, they accumulated, growing over the centuries, over the thousands of centuries, over the millennia, growing over literally hundreds of millions of years on a totally dead earth into a layer sufficient to blanket the entire world three feet deep. Washed and herded by the rains as water vapor erupted out of the earth to form torrential thunderstorms, they would have been driven down into the oceans where, with nothing to destroy them, no predators to eat them, and no oxygen to burn them up, they thickened those primitive oceans into a rich broth approximating a good bowl of chicken soup.

This creation of organic molecules is no mystery. Not only did Urey and Miller demonstrate their creation in the laboratory, but we have seen them appearing abiogenically throughout the universe. The carbonaceous chondrites, which were never part of any planet large enough to support life, contain several of the twenty-some amino acids basic to life, obviously formed naturally by nonliving processes. The interstellar dust is composed not only of inorganic compounds but of such stuff as formaldehyde, hydrogen cyanide, and cyanoacetylene, basic building blocks of the more complicated and necessary biological compounds sugars, amino acids, and pyrmidine bases. It seems that wherever conditions are right, organic compounds will form; and the right conditions are simple enough—an absence of abundant free oxygen, the presence of universally abundant compounds like water, hydrogen, methane, and ammonia, and a source of energy like lightning, ultraviolet light, heat, or static electricity. All of this was present on the early earth, and in addition organics would have fallen in with the meteorites and interstellar dust. Certainly there was no lack of the materials necessary to make life, but it is not sufficient to throw together a stack of wood, nails, hammer, and paint, and expect a house to materialize. Organization is necessary.

Two particular aspects of the organization necessary for the creation of life have been recognized. The first is concentration: although simple amino acids are seen to be spontaneously generated throughout nature, the more complex organic polymers, proteins and nucleic acids, are not. These giant molecules have been created in our laboratories, however, and the difference between the labs and the outside world is simply concentration: put enough of the building blocks together under the proper conditions and the complex molecules will grow. But how are they to be concentrated without some cosmic chemist reaching out his hand and pouring them carefully into his beaker?

By a variety of possible, naturally occurring processes, as it turns out. As water begins to freeze, for example, the dissolved organics do not; instead they concentrate in the remaining solution. At the opposite extreme, as water is heated and begins to evaporate, again the organics do not but instead remain behind in a warm concentrated pool. Inorganic minerals provide another possibility. Because of their particular surface structure some of them are particularly proficient at adsorbing particular molecules and thereby concentrating them on their surfaces—in effect they have prongs or hooks reaching out on which the organics can be impaled and caught. The structure of the hooks is even such that the giant polymerized organic molecules will be caught more efficiently, and thus individually separated and concentrated. The minerals have another pleasing characteristic: they are imperfect, having occasional empty positions in their crystal lattices and occasionally allowing the wrong ion with the wrong electrical charge to take a place, with the result that they generally contain a charge imbalance. These two characteristics, high surface concentrations of organics and a net electrical charge, make them ideal catalysts for further chemical (and eventually biological) reactions.

This first aspect of the creation of life, concentration of the available organic chemicals, is easy enough to accomplish, but the second is more difficult. It arises from the complexity of life and the immutable laws of statistical probability. What we are suggesting for the origin of life is that the naturally formed organic chemicals be brought into random contact with each other, undergo their natural chemical reactions, and little by little the more complex molecules will be formed, until eventually the complexity will reach the level of replication, isolation, and energy withdrawal from the environment, at which stage we say the chemical complex is alive. The problem is that each successive stage must take some finite time, and is not guided but is rather a random event governed by the laws of statistical probability—and probabilities are multiplicative.

Suppose, for example, that you flip a coin: the probability that it will land on heads is $\frac{1}{2}$. If you require that it must land heads twice in a row, the probability is $\frac{1}{2} \times \frac{1}{2} = \frac{1}{4}$: only one time out of four will your requirement be satisfied. Suppose you want to see ten heads in a row: the probability is now only $(\frac{1}{2})^{10} = 0.00098$; flipping the coin once a second, you would have to stand there flipping it for a couple of hours before you'd have a reasonable chance of getting the desired result. (Of course it might happen on any given trial; it might happen the first time you try it, but one would then have to consider the possibilities of a miracle or a fraud rather than simple probabilities.) The time required increases fantastically as the complexity of the required solution increases: for heads to come up thirty-five times in a row the probability decreases to one in a billion, and the experiment would take nearly a thousand years instead of a couple of hours. For the generation of life on earth, the time available is on the order of hundreds of millions of years; but the complexity is so high that simple random processes could not produce the required compounds in times very much greater than that—in fact, the entire history of the universe is nowhere near long enough.

It appears necessary to invoke some kind of guidelines for the life process to begin, some method of selecting the necessary steps as they occur, and preserving them while waiting for the next steps along the road to life. This will remove the damning multiplicity factor from the terrible probabilities, and thus reduce the necessary time immensely. For example, if it is necessary to undergo two different improbable events in sequence, each of which has a probability of occurrence of one in a million, the probability of both happening is $1/1,000,000 \times 1/1,000,000 = 1/1,000,000,000,000$. But if we wait for the first event to happen and then preserve it while waiting for the second event, the total probability is simply $1/2,000,000$. We have increased the probability by a factor of 500,000. A famous analogy, first proposed by the British biologist A. G. Cairns-Smith in 1971, goes back to the old idea of a monkey pounding on a typewriter, hitting keys at random and eventually producing the Bible. If the keys are struck at the rate of one per second, it would take much longer than the entire age of the universe to produce one simple particular sentence; if instead each successive letter were selected and saved until the next could be produced, it would take only a couple of hours.

But such a selection implies that we, or the selecting mechanism, knows in advance what the sentence is to be, or knows what processes are going to be necessary for the formation of life *before life itself has occurred*. Once life is formed and is operating, natural selection is a reasonable such process, selecting inherited traits as a response to environment, but how can chemical formations prior to life be selected? What would be the criterion? The answer is not clear. For an organism to be "alive" it needs both proteins and nucleic acids in at least some stage of its life cycle. Nucleic acids can replicate themselves and thus might choose among succeeding evolutionary paths, but nucleic acids don't do anything except replicate and organize amino acids into protein, so what standards could they have for choosing? It is the proteins that do the work of life, and these in principle would be able to choose between more and less efficient chemical configurations, but proteins do not replicate and therefore don't have the ability to choose. From what we see of the chemistry of biology today, both proteins and nucleic acids together as living units would be necessary to initiate the selection process—but they could not have come together as living units without the selection process. We have, therefore, a little problem.

The solution is not known today, but the puzzle does seem solvable in principle: this conclusion is analogous to recognizing that a particular equation in one unknown is theoretically solvable although it may be so complicated that its solution is not immediately obvious. The key in this situation is likely to be "preadaptation"—selection of traits that will be useful in the future for some other process occurring in the present. The concept seems at first glance to involve some sort of predetermination or parapsychological crystal ball, but in fact does not. It has been used to explain other puzzles in evolution, and may best be understood by considering an example such as adeno-associated virus, AAV.

AAV is a small, single stranded, DNA virus, incomplete in the sense that it is unable to reproduce unless a helper virus is present. If the helper virus is not present AAV remains integrated in the host cell, to emerge only when the helper virus (either adenovirus or herpesvirus) infects that cell. The relationship between AAV and its helper would seem to be a simple symbiotic one, except that AAV inhibits rather than stimulates the growth of its helper: the relationship is beneficial on one side only, and destructive on the other. Any cellular mechanism which slows down the growth of a pathogenic virus such as adenovirus or herpesvirus is beneficial to the host cell but destructive to the invading virus. The slowed-down virus would be analogous to an attenuated virus such as is given in some vaccination procedures, allowing the host defense systems—the inteferon and antibody responses—to be mounted with minimum viral spreading and damage. It is easy to see how symbiotic organisms might evolve a system of mutual dependency under slowly increasing environmental pressures, but how could the antagonistic/dependent situation of the AAV and its helper have evolved?

The solution, suggested in 1985 by Ronald E. Fisher and Heather D. Mayor of Baylor College of Medicine, may be that the evolutionary origin of AAV lies in a cellular anti-viral defense mechanism, beginning with a sequence of "junk" DNA. It is well known that a large portion of the eucaryotic genome does not code for any purpose. This junk or selfish DNA spreads itself throughout the genome but does nothing except rep-licate itself. It can be understood by the idea that each genome is a jungle, and any gene that successfully duplicates itself without actually harming the host will eventually spread even though it serves no useful function. Fisher and Mayor, building on earlier ideas which suggest that such junk DNA might later acquire a useful function, suggest this as a solution to the AAV puzzle. If a particular strand of junk DNA proves capable of interfering with viral replication by competing with the viral DNA for the host cell's replicative enzymes, it is then in effect used by the cell as an anti-viral weapon. This provides a "useful" pressure for its evolution into an element which will compete with the invading virus for necessary chemicals within the host cell.

Once the junk DNA evolves this far, a further period of evolution would take place in which the primary selection pressure would be toward its own stability within the cell. Although no intelligence is involved, the evolutionary behavior of the AAV is as if it were seeking to find its own niche in life rather than simply being used as a tool by the host cell. In this sense, then, if it fails to compete effectively with the invading virus it will be destroyed; on the other hand if it is too successful the invading virus will be destroyed and the AAV will be reintegrated into the host cell and never emerge again into its own "personality." It therefore faces the same dilemma facing the military forces in our own civilization, and has reached the same solution, evolving to a point where it is effective enough to prevent the invading virus from killing the host cell, but not efficient enough to destroy the invading virus completely; in this way its continuing presence is necessary—it has found its niche.

In a similar manner life itself may have formed on a prebiotic earth, with function following rather than dictating form. Particular organic chemicals will have evolved in a random, "nonintelligent" manner when energy such as lightning flashes passed through simpler inorganic compounds naturally present; they may have been selected for their structure in fitting into or onto particular mineral species (perhaps helping them to withstand destruction by weathering processes and thus forming a nonliving, nonintelligent form of symbiosis). Eventually these organic compounds replicated, sealed themselves off from the environment by the first forms of a cell membrane, and became "alive." The future course of life on earth was in this sense predetermined by the usefulness of some specific organic molecules for a purpose which has no link with present life forms.

Granted that something of the kind occurred, what are the implications for life elsewhere? First, we note that the simplest organic building blocks of life are precisely those that are formed most easily via inorganic pathways, i.e., that are formed throughout the universe. Second, the future organization of these molecules depends on local conditions unrelated to future life forms. The conclusion is that life anywhere in the universe where conditions are suitable will develop along basic lines of construction similar to our own, but that the particular architecture will be unique to its own environment. The one thing that we can say for sure is that life everywhere will be based on carbon and the compounds it makes with hydrogen, nitrogen, and oxygen.

This last point has been emphasized by, among others, George Wald of Harvard. Carbon is the only possible starting point for the molecules of life, which must have two characteristics: the capability of complexity, and stability against chemical attack. Complexity is obtained via carbon's ability to form long, virtually endless chains of atoms and by its ability to react with different atoms by forming a variety of bond types: single, double, or triple, depending on its partner. It is sometimes suggested in science fiction magazines that such thinking is too anthropomorphic, too chauvinistic, and that elsewhere in the universe life might be completely different, choosing as its basis silicon instead of carbon.

Silicon is suggested because its atomic structure is similar to carbon, and thus it is chemically similar and similarly able to form complex molecules. But its structure, though similar to carbon, is not identical and results in two deficiencies when considered as a basis for life: the complex molecules it forms are less tightly bound and therefore less stable, and the atomic structure includes unfilled inner electron orbitals which are available for chemical attack. The result is that such silicon-based "organic" molecules could not long exist in an environment containing such active "aggressor" compounds as water, ammonia, or oxygen. (It might be noted that on earth silicon is more than a hundred times as abundant as carbon. If it were possible to base life on it rather than on carbon, this simple consideration should have been enough to have made us all silicon creatures. But we are not. This fact alone demonstrates that the inner virtues of carbon are overwhelming.)

So carbon as a basis is necessary; the reason for its combination with hydrogen, oxygen, and nitrogen is that these are the smallest atoms in the universe capable of forming stable compounds with carbon—and the small size of the resulting bonds makes them stronger than any others. And that's all there is to it. In reply to the hypothetical possibility that there might exist elsewhere in the universe some new element unknown on earth, that would be as suitable as carbon, Professor Wald has written: "I have been asked sometimes how one can be sure that elsewhere in the universe there may not be futher elements other than those in the Periodic System (i.e., unknown to us here on earth). I have tried to answer by saying that it is like asking how one knows that elsewhere in the universe there may not be another whole number between 4 and 5. Unfortunately, some persons think that is a good question, too."

Life developing elsewhere in the universe (or starting over again here on earth if we destroy it all and leave the planet intact) would therefore be constrained to follow identical initial pathways, but would very soon encounter nearly infinitely diverse branches along which it could travel. Each species of life on earth is a vastly improbable result of successive choices among possibilities too numerous to count, any one of which would have led to a different final result if a different evolutionary choice had been made. So we must not expect to find, on some planet circling some sun in some galaxy on the other side of beyond, human beings or cockroaches or roses or staphylococci. But what might we expect?

First, it must be emphasized that we are speaking from a data set comprised of one datum: we know of precisely one planet on which life has evolved. (The *Viking* rockets sought life on Mars and returned some data which were inconclusive, though the consensus interpretation is that the planet is barren; certainly there is no Martian data which can indicate to us the type of life which might have existed there, if ever it did.) If we were conducting a properly controlled scientific investigation, it would be time now to gather more experimental data rather than to speculate with what we have. But the Great Cosmic NSF in the sky doesn't give us what we ask for any more easily than does the one in Washington, so we have to take what we get; and what we get today is speculation based on our one datum.

The prime example of how a mathematical equation can express a quantitative relationship between a physical concept and pure fantasy (given our present state of ignorance) is the one formulated by Frank Drake, a radioastronomer then at Arecibo Observatory and currently dean of Natural Sciences at Santa Clara. It defines the number of civilizations *(N)* in our galaxy as:

$$N = R \times P \times E \times L \times I \times C \times T$$

where, reading from left to right, we pass from semiquantitatively known functions to astropoetry. (The suffix form "-poetry" is a pejorative one among scientists, though the noun itself is treated with affection and respect.) *R* is the average rate of formation of stars in the galaxy, and sufficient data exist to permit a reasonable estimate of about ten stars formed per year, averaged over the lifetime of the galaxy. Obviously there is a large error here, since the data were all obtained recently and must be extrapolated to distant times, but it's probably good within an order of magnitude. The next function takes us over the border from semiquantitative reasoning to theory, imagination, and hope: *P* is the fraction of stars that have planets, and of course is the fulcrum upon which the arguments of this book swing and sway. It is the contention of all nebular theories of planetary formation that the process is a common one, so that if we are somewhere near correct the function *P* is somewhere near unity. *E* is the environmental function, the fraction of planets that have an environment suitable for the development of life. For a suitable environment, there must first of all be present the necessary chemical elements, including carbon, nitrogen, and oxygen, as well as hydrogen. The first three of these are formed only in the interior of stars in advanced stages of development, and are dispersed into the interstellar medium only when these no-longer-young stars belch away their outer envelopes. Therefore the first stars to form in the galaxy, together with any planets that might form with them, will consist of nothing but hydrogen and helium—so no possibility of life on these. But even for the later planets, with sufficient carbon and other elements, there are many other necessary environmental factors. The temperature must lie within the boiling and freezing points of water. In our solar system, for example, Mercury and Venus are too close to the sun and therefore too hot for life, while the planets further away than Earth are probably too cold. Planets much smaller than Earth could not hold sufficient atmosphere to allow the existence of liquid water and to moderate temperature variations between night and day, while planets much bigger might retain an atmosphere so thick that sunlight could not penetrate it. The restrictions on the possible range of planetary conditions are strict.

These first three factors together give the probable number of planets in our galaxy suitable for the possibility of life, and numerical estimates run from essentially 0 to 10 billion! Obviously we are in need of some serious hard data, although there is a growing consensus that while it may be difficult to produce a satisfactory quantitative estimate, the probable number is more likely to be nearer the high end of this range than the low.

The next factor, L, is the fraction of suitable planets on which life actually develops. The general feeling is that this factor is essentially unity: what can happen, will happen; if life can form, it will form. We don't really know this, the conclusion is closer to faith than pragmatism, but at least there is nothing in all we have learned about the chemistry of life to indicate the existence of an insurmountable barrier to its formation, except the rather precise conditions that must be met (contained in the factor E).

These functions describe the number of planets on which life is expected to exist. The next one, I, describes the probability of intelligent life. Since our understanding of the nature of intelligence is practically nil, it might seem that this function is totally unknown; but actually we can say a bit about it. We know, for example, from our own experience that intelligence is hardly the first development of a living system. On earth it took about four billion years for life to develop a consciousness of its own existence, which is the first requirement of intelligence. Many stars, those more massive than the sun, burn hotter and more fiercely; they use up their fuel too quickly, racing through their life cycles without sufficient time for intelligent life to evolve. When our own sun passes its main sequence stage and swells into a red giant, it will expand until its outer envelope encloses the earth, raising temperatures here to thousands of degrees and ending all life on this planet. Hopefully by that time we will be able to pack up and go elsewhere (we have another five billion years or so to make our plans), but if our sun were more massive and correspondingly shorter-lived, if it were to have reached that stage a billion years ago, for example, all life here would have perished before it ever had a chance to go intelligent. So intelligent life is probably restricted to the smaller stars like our own, which live for at least a few billion years.

The penultimate factor C is the fraction of such intelligent forms that reach the stage we call civilization. Here again you pays your money and you takes your choice. You may feel, as I do, that intelligence given sufficient time must reach its own form of civilization, or you may feel that civilization is a unique outreach of our own unique species. The only possible way to settle this question is to search for other civilizations on other worlds. If we find one, the argument ends with a bang; if we do not, the argument dwindles down to a whimper as star by star is shown to be devoid, and finally we will have to admit that there is no one out there.

While searching, it's amusing to consider the last factor, T, the mean lifetime of civilizations. Are we, as a society, immortal? Or will we one day destroy ourselves? Might we not fall victim to the same unknown causes that destroyed the dinosaurs and countless millions of other species on this planet throughout geological time? Bubonic plague nearly destroyed European civilization, and our recent replies to herpes and AIDS give no support to the comfortable delusion that medical science is sufficiently efficient to meet any sudden new pathological threat.

With this in mind, let's discuss for a bit some realistic attempts to contact whatever civilizations may be out there in the far reaches of the galaxy.

First of all, we recognize that we have to reach out beyond our own solar system and that we are constrained to stay within our own galaxy. In our own system there may yet be life on Mars, Jupiter, or Titan (Saturn's giant moon), to name the most likely possibilities—but all of these most likely locales are still very unlikely. And certainly if life has formed in any of these places it will consist of very primitive forms, which doesn't make the possibility an uninteresting one, but certainly we don't expect Martian microbes to send us intelligible messages. So we must reach out beyond this local planetary system we are embedded in. At the same time we are constrained to search only within our own galaxy, at least during the foreseeable future, simply because of the vast distances involved. The nearest galaxy beyond our own is that of the Magellanic Clouds, which lie about 200,000 light-years away. Any communication, of necessity limited to the speed of light, would involve this time delay between signal propagation and reception, diminishing interest in initiating any such attempt. In addition, the intensity or probability of reception of any signal must fade with the square of the distance, and so the difficulty and expense must increase with it, and so the immense intergalactic distances argue strongly for looking closer to home, at least at first.

Even within our own galaxy the distances are vast and daunting. The nearest stars are light-years away, and there aren't more than a few within several dozens of light-years. If the fraction of stars with advanced civilizations is small, we are likely to have to search many random stars before we find an intelligible voice, and the distances involved quickly become hundreds and thousands of light-years. To send a spacecraft out to each of these is obviously a hopeless endeavor.

How then can we reach out to the rest of the universe? The only possible hope is by radio waves, which travel at the speed of light and spread out spherically to reach the stars in all directions simultaneously. But we are not quite at the stage in our development where we can spare the money to build the necessary high-powered transmitters. Happily, there is one alternative because of our young age.

Our world is four and a half billion years old, our technological civilization barely a hundred years old. The first stars in our galaxy formed about 10 billion years ago, and though these did not have the heavy elements necessary for solid planets and thriving life there has been plenty of time for the first massive planets to run through their life cycles and erupt their stellar-synthesized elements out into space, and for second generation stars to form millions and billions of years before the sun. If life did originate on any of these, it would have had all this time to develop before our own planet formed, and today its civilization might be millions or billions of years more advanced than our own. When we consider how our present civilization compares to those of just a couple of thousand years ago here on earth, the thought is staggering.

If such advanced civilizations do exist, it is possible that they share with us an innate desire to search out other life and communicate with it. Certainly with us this desire is instinctive. No sooner had our ancestors learned to dig out a canoe from a fallen tree than they pulled it into the water and set off to see what lay beyond the horizons. From the islands of the Pacific to the Sea of Japan, from the Great Lakes to the Atlantic Ocean, every primitive group of humans on earth went sailing across the waters. If other beings exist on other worlds, they may well share the same instinct. The assumed ubiquity of this instinct, in fact, has been used to argue against the possibility of advanced civilizations elsewhere. For if they do exist, by the same argument given above some of them must be more advanced than we, and where are they? Why haven't they contacted us? Why aren't flying saucers real?

Some suggested answers to this puzzle of our solitude reach fanciful heights, such as that we may be living in a zoo kept by such an advanced race, unaware of the invisible bars that keep us isolated and natural in our primitive state for the weekend enjoyment and serious anthropological study of galactic citizens; but a more realistic view is simply that space is so vast that personal exploration *a la* the *Nina, Pinta,* and *Santa Maria* is out of the question, to be replaced by radio exploration. The idea is that an advanced civilization that wants to communicate with others would send out radio signals indicating their presence. The first contact would come when someone replied, and perhaps eventually more material contact would be affected. But certainly no one could afford to send space craft searching at random throughout the tremendous and empty spaces of our galaxy.

The answer, then, to the problem of how to contact the aliens is simply to listen.

This was first done in 1924 when Mars made a close approach to Earth (since their orbits are elliptical and the two planets travel at different speeds, the distance between them continually varies between maximum and minimum). This was at a time when many people still believed that the canals of Mars existed and were proof of a superior planet-wide civilization there. A retired professor from Amherst College suggested that all our radio stations should shut down at a predetermined time and that we should all listen to see if we could hear any transmissions from Mars. We did, and of course we didn't.

But in 1959 Phil Morrison and Guiseppe Cocconi of Cornell published a short paper in the British journal *Nature,* suggesting that radio astronomy had progressed to the point that we might be able to detect radio signals from any galactic neighbors that might exist. They even suggested which station we should tune into. This was vital, for radio telescopes, like normal radio receivers, must accept a limited portion of the available wave length spectrum at a time. If we search at one wavelength, or frequency, but our friends are broadcasting on another, we'd never find them. Morrison and Cocconi pointed out that there exists a natural choice, "a unique, objective standard of frequency which must be known to every observer in the universe." Hydrogen naturally emits radiation with a wavelength of 21 centimeters, and since hydrogen is the most abundant element in the universe this becomes a natural choice. Other scientists have suggested the frequency of the hydroxyl ion (OH^-), and today the band of frequencies bound by these two is called the "waterhole," since $H^+ + OH^- = H_2O$ and since the waterhole is the traditional and most natural place for wandering tribes to meet.

In 1960 Frank Drake, who was then at the National Radio Observatory in Virginia, began just such a search. Limited in scope by virtually zero funding, his Project Ozma listened to two nearby stars for several months. It heard nothing; nothing, that is, that could be distinguished from the natural cosmic microwave radiation that is the basis for the science of radio astronomy.

And then in 1967 a graduate student at the University of Cambridge, Jocelyn Bell (now Jocelyn Burnell) found an extremely peculiar behavior in the radio transmissions from one particular star. Instead of being continuous, as they must be from any natural source, the radio emissions were coming in precise bursts, repeated extremely regularly at intervals of 1.3373013 seconds. This was so unnatural that members of the radio observatory staff thought they had actually detected signals from a civilization out there. But when further searches were made, similar signals were found to be coming from literally thousands of sources; the magnitude of supposed civilizations was too much, an embarrassment of riches, and the explanation was found in a new type of star, a rapidly spinning neutron star with an intense magnetic field which constrains its radio emissions like the beam of a searchlight so that we detect it only when it flashes by us.

Since then the Search for Extraterrestrial Intelligence has had its own acronym (SETI) and little else, but it has stayed alive, possibly only because it is taken seriously in Russia and we mustn't take a chance on them making contact first. It has stayed alive almost entirely within the minds of scientists, which you might think is not a bad environment for a scientific idea, but scientific ideas, like the AAV virus, need a "helper" to live and flourish: they need an infestation of money, and Congress has repeatedly refused to allow NASA to spend any money at all on this program. Generally the funding agencies such as NASA are specifically empowered to make decisions as to what scientific proposals to fund, based on review of the merits of each proposal, but in 1981, when NASA specifically listed SETI in its request for that year's funds, Congress passed a specific amendment to the budget forbidding the expenditure. It was an unfortunate act, if only because the decision in Congress was made obviously on political rather than scientific grounds: William Proxmire had given SETI one of his Golden Fleece awards a few years previously, citing it in his wisdom as a waste of taxpayers' money, and the political influence and newspaper coverage of the award is enough to frighten people away (especially people who don't much care anyway). But in 1983, after years of lobbying by organizations like the Planetary Society, a semiscientific organization devoted to cosmic exploration, and by well-known individuals like Carl Sagan, and by the scientific community, Proxmire backed off and Congress did allocate SETI money for NASA: $1.5 million a year for five years.

The prospects of success have been multiplied tremendously recently not only by the availability of money but by the perfection of the multi-channel spectrum analyzer, which allows a single radio telescope to listen on millions of frequencies at the same time instead of picking out just one restricted band. A search as intense as that carried out by Frank Drake in his several-months-long Ozma experiment would take only a few seconds now; a corresponding amount of time would be millions of times more effective. With it, the SETI attempt has reached a level of probability that removes it from the level of pipe dreams and deposits it in the realm of distinct possibilities. Both the National Academy of Sciences and the International Astronomical Union have joined with the Planetary Society in giving moral support and encouragement.

The experimental plan has been argued over, and at least the first phase has been decided upon. It will be a two-pronged attack, first sweeping the entire sky, listening to each section for only a second or two, and then picking out several hundred of the closest and most likely stars and concentrating on each of them for several minutes. This represents a compromise between two basically opposing strategies: picking out the most likely astronomical places for life to exist, or just looking everywhere. The former has the psychological advantage to the searchers of feeling more in control of their experiment, but the slight disadvantage of impossibility. The stars that have been seen to have planetary disks are all young stars, their planetary system just forming, too young to have developed life, let alone intelligence. Planetary material lumped into a few discrete planets is much harder to detect than its predecessor, the nebular disk, and so the stars with actual planets are still unknown to us.

The other extreme line of attack, simply looking all over and hoping, leads to the disturbing possbility that another civilization with its own budgetary restrictions may be transmitting only on an eight-hour day, and at the moment we happen to scan over them they may be silent. More importantly, our hopes must be daunted by the size of the jungle out there, and the expected smallness of distribution of intelligent life. Going back to Drake's equation, while there may be a total of millions or even billions of inhabited planets, the fraction of any of these having successively intelligent, civilized, and finally advanced communities drops quickly down, so that even if there are many such civilizations throughout the galaxy, there are infinitely more places where they are not.

There are other problems as well. At our own advanced stage of civilization, it is clear to us that exploration of the galaxy by physical means, spaceships and the like, is hopelessly inadequate, and so we turn to radio waves which have the advantages of propagating at the maximum speed allowed by Einstein's theory of relativity, the speed of light, and of spreading out in all directions simultaneously and continuously. No physical craft can match this. But suppose there is something else that is even better. What? I don't know, no one knows, and that's the whole point. If we go back in our history just over a hundred years, no one knew about radio waves. People before the radio era had suggested that we light giant bonfires to signal our supposed brothers on Mars, or build immense geometrical structures out in the plains of Kansas or Siberia so that any passing spaceships would see them and perhaps reply to our hello. These people were not crackpots; one of them was Karl Friedrich Gauss, one of the greatest mathematicians who ever lived. And the ideas weren't stupid, merely naive—and naivety is a relative term. Perhaps in our reliance on radio waves we are being incredibly naive, while out there a galactic wave of civilizations is talking to each other by means of which we know nothing.

Perhaps. But all we can do is work with what we have. So we'll search the radio wave spectrum and look for—what? How does one establish contact across not only the immensity of space but across the limitless immensities of totally different possible life forms? What can be said that would be recognizable as a distinct signal, a cry of "Hello, out there!" When the *Pioneer* spacecraft left our solar system and started their long aimless wandering through interstellar space, they carried within them plaques depicting a naked man and woman; the *Voyager* spacecraft, which by the twenty-first century will have left Pluto far behind to begin their own wandering, carry a more complicated message: a 12-inch gold-plated copper phonograph record containing among other things a Bach fugue and a Beethoven string quartet, so that if anyone should ever find it they would have an idea of what we are like. But of course if anyone ever found it they would know more about us from the spacecraft itself and its instruments' construction and level of sophistication than from a picture of us naked, or from the sounds of Beethoven and Bach. . . . Or perhaps not. It's something to think about. At any rate they would know that the spacecraft was not a natural rock, that it had been purposely created.

The simple receipt of a radio signal that was clearly not natural in origin would have the same impact, but might not be easy to discern. A repeated message would of course be obvious; say, one dot followed by two followed by three, and then repeated. But in expecting something like this are we not expecting something anthropomorphic in thought pattern? It's so hard to get away from this because we are, after all, the only examples of intelligent life known to us. It's difficult to predict the natural behavior of intelligences arising from different biochemical reactions, evolving in totally different environments for billions of years. Another way to look for signals is to search for a transmission on one frequency of much greater intensity than seen on its neighboring frequencies, since we know of no natural processes that transmit in this fashion. On the other hand, we also thought that natural processes didn't transmit in pulses until we discovered pulsars, so any "intelligent" signal detected in the future is going to be put to a lot of scrutiny before it's accepted for an intragalactic hello.

And after that it gets even more difficult. How to decipher such a signal, how to wring out of it any meaning? The one thing we know about the universe is that it is put together according to mathematical relation-ships, and so mathematicians have long amused themselves with attempting to put together a truly universal mathematical "language," but none have succeeded. Even so, surely the most important result in the history of human exploration would be the determination that a received signal from somewhere out there was purposeful, intelligent in origin. It would have to affect everything we do from then on, every form of intelligent human endeavor: our philosophy, our religion, our self-knowledge and self-image. Few accomplishments open to us would have the same impact on our future behavior as this: simply to establish the synthetic, intelligent origin of a purposeful radio signal against the continuous background of natural cosmic radio noise, to establish that we are not alone.

Or, if the search continues without any positive result, to show that we are alone.

Either way, it would be nice to know for sure.

SOURCES: Fisher and Mayor 1986; Papagiannis 1985

APOLOGIA

Tennyson once said of Browning's *Sordello* that he understood only the first line

> Who will, may hear Sordello's story told.

and the last

> Who would, has heard Sordello's story told.

and that both were lies. I have tried in this book to make all the lines understandable, but necessarily some of them are lies—or at least they are not the whole truth in the sense of telling a whole and complete story. Some of this is unavoidable because of the nature of the beast: we cannot expect any story of the origin of the earth and planets to explain everything about them, any more than the story of Hitler's birth would explain all the inconsistencies of his later life. For both people and planets, much of lasting importance happens well after birth.

The retrograde motion of Venus, for instance, together with that of several of the moons of the Jovian planets, has not been explained and cannot be as part of their origin. No reasonable story of their birth can account for their spinning or orbiting backwards. This must have happened afterward—or at least toward the end of their creation process. For example, a nearly formed Venus may have been hit in a large glancing impact sufficiently energetic to spin the planet around in reverse. In the case of the outer Jovian moons the story is probably one of capture of bodies initially extrinsic to the Jovian system; such capture could lead equally well to prograde or retrograde orbits, and so if all the outer moons were captured it is reasonable that a fair percentage exhibit retrograde motion. In fact one would expect all the moons to have prograde motion only if all the accreting bodies had low orbital eccentricities and were small compared to those on which they accreted, which is not necessarily the case. But I haven't gone into any of this in detail because it's beyond both the scope of this book and our present knowledge.

The Titius-Bode law also has not been explained. In this case it is obviously related to the origins of the planets, but it is not yet clear that it actually exists as a physical consequence of mathematical relationships in a condensing or aggregrating protosolar disk, rather than as merely a numerologic curiosity (with numerology bearing the same relationship to mathematics as astrology does to astronomy). Arguments can be made on both sides of the question.

A more important lie in this book is the omission of several lines of argument that currently lie beyond the bounds of what we may reasonably claim to understand, but if scientists were to wait for all such arguments to be settled before writing on any subject, no scientific books would ever be written at all. Oliver Manuel, for example, at the University of Missouri, believes that the sun and planets, instead of forming from a cloud of interstellar gas and dust, condensed directly from the remains of a supernova, with the sun forming directly on the core, the inner planets forming from the inner regions and the outer planets from the outer regions, with some mixing due to the sun pulling in some outer region material to accrete on the inner planets. In this scheme the sun is a totally different creature from what it is believed to be by nearly everyone else, being composed mostly of iron rather than hydrogen and helium, and depending on something other than fusion for its energy source. It is the consensus that this model is not likely to be valid, for too many physical and mechanical reasons to discuss here, and so with apologies I have left it alone.

There are always such models floating around at conferences and *kaffeeklatsches*, and while they are interesting for specialists to argue about they have no place in a book like this. In a different class is the argument put forward by Tommy Gold, one of the most scintillating, irritating, and brilliant astrophysicists of this century. Among his brilliant suggestions which turned out wrong were those which predicted that the surface of the moon might be dust thick and soft enough for our spaceships to sink into and disappear (which threw the early Apollo program into a panic) and that the universe itself might never have been created but rather existed always through eternal time. Among his ideas which turned out to be right is the identification of pulsars with the theoretically predicted neutron stars, and a host of details about the structure of the solar system and the moon.

In a speech given last year to the British Association of Science he brought up the previously conveniently overlooked inconvenient angle of 7.2° between the plane of the equator of the sun and the plane of total angular momentum of the planetary system. If the nebular disk and subsequently the planets formed by magnetic braking and angular momentum dissipation, these two planes should be coincident. Certainly there must be some relationship between the spin of the sun and the motion of the planets, for if they were totally unrelated the angle of 7.2° is too small: the chance of two vectors without some causal interaction falling accidently within such a small angle is only one in four thousand, and for a direct cause-and-effect mechanism such as magnetic braking of a protosolar disk the angle is too large: it should result in precisely synchronous planes of revolution.

Furthermore, the current model of planetary formation envisages a cloud collapsing toward the center, spinning rapidly and then decelerating as its magnetic field sends angular momentum outward. But this results in mass also moving outward: we would have a nebular disk first contracting, then expanding, and this certainly implies a good deal of mixing within the nebula. We would expect a homogeneous disk to result. Such homogeneity was indeed the initial conclusion of meteoritic workers, who found a high level of isotopic sameness wherever they looked in solar system materials. But the more recent results discussed in chapter 23 established that the material out of which the solar system formed was not totally homogeneous. It is not at all clear at the present time how to keep a magnetically braked nebular disk from achieving isotopic homogeneity.

These sorts of problems are real and important, and must be answered before a complete theory of planetary formation is accepted. But they are qualitatively different from the situation at the beginning of the century, when the angular momentum problem was an absolute one: unsolvable in principle. These problems are (or seem to be, one hopes) the sort of problems that science is meant to deal with, the sort that are in principle amenable to further investigation.

Up until the last few decades, those who were searching out the secrets of our creation were wandering lost in a forest, threshing about wildly with no idea of where their destination might be or what form it might take, plunging forward in one direction for a generation, then suddenly reversing and pushing wildly off at right angles, not knowing when they were making progress or when they were regressing. Now at last we see the light beckoning through the gloaming, we see the shape of the structure taking place as we peer through the dead branches—but we've still miles to go before we sleep.

Ah, but a man's reach should exceed his grasp,
Or what's a heaven for?

—Robert Browning

Nothing worth knowing can be understood.

—Woody Allen

BIBLIOGRAPHY

(A complete bibliography would run to many hundreds of citations. The nature of the following, which were particularly helpful to me and/or which provide useful further reading, range from the popular literature through textbooks to research articles, as should be clear from the titles.)

Barnes, C. A., D. D. Clayton, and D. N. Schramm, eds. 1982. *Essays in Nuclear Astrophysics* Cambridge: Cambridge University Press.

Brown, Peter Lancaster. 1974. *Comets, Meteorites, and Men.* New York: Taplinger.

Black, David C., ed. 1984. *The Planetary Report,* September/October.

Black, D. C. and M. S. Mathews. 1985. *Protostars and Planets II.* Tucson: University of Arizona Press.

Boss, Alan P. 1985. "Collapse and Formation of Stars." *Scientific American,* January.

Brown, G. C. and A. E. Mussett. 1981. *The Inaccessible Earth.* London: Allen and Unwin.

Brownlow, Arthur H. 1979. *Geochemistry.* Englewood Cliffs, N. J.: Prentice-Hall.

Brush, Stephen G. 1980. "Discovery of the Earth's Core." *American Journal of Physics,* 48:705.

Brush, S. G. 1981. "From Bump to Clump: Theories of the Origin of the solar system 1900-1960. In P. A. Hanle and V. D. Chamberlain, eds., Space Science Comes of Age. Washington, D.C.: Smithsonian Institution Press.

Brush, Stephen G. 1982. "Chemical History of the Earth's Core." *EOS* 63:1185.

Burbidge, E. M. and G. R. Burbidge. 1982. "Nucleosynethesis in Galaxies," in Barnes et al., eds., *Essays in Nuclear Astrophysics*.

Burbidge, E. M., G. R. Burbidge, W. A. Fowler, and F. Hoyle. 1957. "Synthesis of the Elements in Stars." *Reviews of Modern Physics*, 29:547.

Burchfield, Joe D. 1975. *Lord Kelvin and the Age of the Earth*. New York: Science History Publications.

Cameron, A. G. W. 1985. "Formation and Evolution of the Primitive Solar Nebula." London: *Philosophical Transaction of the Royal Society of London* (A). 313:5.

Charon, Jean. *Cosmology*. 1970. New York: McGraw Hill.

Clarke, Roy S. Jr., Eugene Jarosewich, Brian Mason, Joseph Nelen, Manuel Gomez, and Jack R. Hyde. 1970 "The Allende, Mexico, Meteorite Shower." *Smithsonian Contributions to the Earth Sciences*, no. 5.

Dermott, S. F., ed. 1978. *The Origin of the Solar System*. New York: Wiley, 1978.

Dobzhansky, Theodosius, Francisco J. Ayala, G. Ledyard Stebbins, and James W. Valentine. 1977. *Evolution*. San Francisco: Freeman.

Faul, Henry. 1978. "A History of Geologic Time." *American Scientist* 66:159.

Faul, Henry and Carol Faul. 1983. *It Began with a Stone*. New York: Wiley.

Field, George B., Gerrit L. Verschurr, and Cyril Ponnamperuma. *Cosmic Evolution*. Boston: Houghton-Mifflin, 1978.

Fisher, David E. 1977. *Creation of the Universe*. Indianapolis: Bobbs-Merrill. 1977.

Fisher, David E. 1979. *Creation of Atoms and Stars*. New York: Holt Rinehart, Winston.

Fisher, Ronald E. and Heather D. Mayor. 1986. "Evolution of a Defective Virus from a Cellular Defense Mechanism." *Journal of Theoretical Biology* 118:395.

Grossman, L. 1972. "Condensation in the Primitive Solar Nebula." *Geochimica Cosmochimica Acta* 36:597.

Habing, Harm J. and Gerry Neugebauer. 1984. "The Infrared Sky." *Scientific American* (November), p. 49.

Hack, Margherita. 1984. "Epsilon Aurigae." *Scientific American* (October), p. 98.

Hemingway, Ernest. 1926. *The Sun Also Rises*. New York: Scribner's.

Herbst, William and George E. Assousa. 1979. "Supernovas and Star Formation." *Scientific American* (August), p. 138.

Hoffman, Banesh. 1972. *Albert Einstein: Creator and Rebel*. New York: Viking.

Hoyle, Fred. 1975. *Astronomy and Cosmology*. San Francisco: Freeman.

Hoyle, F. 1982. "Two Decades of Collaboration with Willy Fowler." In Barnes et al., eds., *Essays in Nuclear Astrophysics*.

Inglis, Stuart J. 1972. *Planets, Stars, and Galaxies*. New York: Wiley.

Kaufmann, William J. III. 1979. *Planets and Moons*. San Francisco: Freeman.

Kerr, Richard A. 1984. "Periodic Impacts and Extinction Reported." *Science* 223:1277–79.

Kerr, Richard A. 1984. "Making the Moon from a Big Splash." *Science* 226: 1060.

Krinov, E. L. 1960. *Principles of Meteoritics*. New York: Pergamon.

Lockyer, J. Norman. 1890. *The Meteoritic Hypothesis*. New York: Macmillan.

Lovell, Bernard. 1981. *Emerging Cosmology*. New York. Columbia University Press.

McBride, Ken. 1984. "Looking for Extrasolar Planets." *Astronomy* (October), 12:8.

Menzel, Donald H., Fred L. Whipple, and Gerard deVaucouleurs. 1970. *Survey of the Universe*. Englewood Cliffs, N.J.: Prentice-Hall.

Morrison, David and Jane Samz. 1980. *Voyage to Jupiter*. Houston: NASA.

Morrison, Samuel P. 1971. *The Discovery of America: The Northern Voyages*.

Murray, Bruce, Michael C. Malin, and Ronald Greeley. 1981. *Earthlike Planets*. San Francisco: Freeman.

Nininger, H. H. 1952. *Out of the Sky*. New York: Dover.

Ozima, Minoru. 1979. *The Earth*. New York: Cambridge University Press.

Paneth, F. A., P. Reasbeck, and K. I. Mayne. 1952. "Helium 3 Content and Age of Meteorites." *Geochimica Cosmochimica Acta* 2:300.

Papagiannis, M. D. 1985. "Recent Progress and Future Plans on the Search for Extraterrestrial Intelligence." *Nature* 318:135.

Payne-Gaposchkin, Cecilia and Katherine Haramundanis. 1970. *Introduction to Astronomy*. Englewood Cliffs, N. J.: Prentice-Hall.

Peale, S. J., P. Cassen, and R. T. Reynolds. 1979. "Melting of Io by Tidal Dissipation." *Science* 203:892, 1979.

Podolak, M. and R. T. Reynolds. 1984. "Consistency Tests of Cosmogonic Theories from Models of Uranus and Neptune." *Icarus* 57:102.

Reynolds, J. H. 1960. "The Age of the Elements." *Physical Review Letters* 4:8.

Ringwood, A. E. 1979. *Origin of the Earth and Moon*. Berlin: Springer-Verlag.

Schramm, D. N. 1982. "The r-process and Nucleocosmochronology." In Barnes et al., eds., *Essays in Nuclear Astrophysics*.

Schreiber, Edward and Orson L. Anderson. 1970. "Properties and Composition of Lunar Materials: Earth Analogies." *Science* 168:1579.

Scoville, Nick and Judith S. Young. 1984. "Molecular Clouds, Star Formation, and Galactic Structure." *Scientific American* (April), p. 42.

Simpson, George Gaylord. 1976. *Penguins*. New Haven: Yale University Press.

Suess, H. and Harold C. Urey. 1956. "Abundances of the Elements." *Reviews of Modern Physics* 28:53.

Stothers, Richard B. 1985. "Terrestrial Record of the Solar System's Oscillation about the Galactic Plane." *Nature* 317:338.

Swihart, Thomas L. 1978. *Journey Through the Universe*. Boston: Houghton Mifflin.

Tarbuck, Edward J. and Frederick K. Lutgens. 1984. *The Earth*. Columbus, Ohio: Merrill.

Wainwright, G. A. 1932. "Iron in Egypt." *Journal of Egyptian Archaeology* 18.1.

Wald, George. 1964. "The Origins of Life." *Proceedings of the National Academy of Sciences* 52:595.

Wasserburg, G. J. and D. A. Papanastassiou. 1982. "Some Short-Lived Nuclides in the Early Solar System: A Connection with the Placental ISM." In Barnes et al., eds., *Essays in Nuclear Astrophysics*.

Weinberg, Steven. 1977. *The First Three Minutes*. New York: Basic Books.

Wetherill, George. 1981. "The Formation of the Earth from Planetesimals." *Scientific American* (June), p. 162.

Wilson, David. 1983. *Rutherford, Simple Genius*. Cambridge: MIT Press.

Wood, John. 1977. "Ancient Chemistry and the Formation of the Planets." In *Proceedings of R. A. Welch Foundation Conferences on Chemical Research, Cosmochemistry*.

Wood, John A. 1968. *Meteorites and the Origin of Planets*. New York: McGraw-Hill.

Wood, John A. 1979. *The Solar System*. Englewood Cliffs, N. J.: Prentice-Hall.

INDEX